IMAGES
of Motoring

MORRIS MINOR
LIGHT COMMERCIALS

Mac Fisheries van, c.1960, fitted with small rear door windows and a single rear lamp unit. Later models from 1963 had larger windows and a twin light unit when the indicator was added. *Photo: B Russell.*

IMAGES
of Motoring

MORRIS MINOR
LIGHT COMMERCIALS

Compiled by
Russell Harvey

TEMPUS

First published 2000
Copyright © Russell Harvey, 2000

Tempus Publishing Limited
The Mill, Brimscombe Port,
Stroud, Gloucestershire, GL5 2QG

ISBN 0 7524 1735 5

Typesetting and origination by
Tempus Publishing Limited
Printed in Great Britain by
Midway Clark Printing, Wiltshire

This book is dedicated to my wife Mandy,
for the many hours that she has endured
as a 'book widow' and to my son James,
for constantly having to 'give up'
the computer to me.

Contents

Introduction		7
1.	Minor LCV (Light Commercial Vehicles)	11
2.	GPO, PO and Kingston Telephones Vans	33
3.	Adapted Minor LCVs and Unique Features	51
4.	Public Utilities (Including Local Councils and Government Departments	67
5.	Police Minor Vans (Including Military and Emergency LCVs)	75
6.	Malta and Cyprus	85
7.	Roadside Assistance	93
8.	Minor LCV Register and Current Vehicles (Including Film and TV)	101
9.	LCVs Overseas	115
Acknowledgements		128

A typical sales advertisement from the late 1950s. It makes reference to the new 950cc engine, which was introduced in 1956. These advertisements would have appeared in the motoring press and journals at that time.

Introduction

The Minor LCV (Light Commercial Vehicle) was produced from 1953-1971, although numerous vehicles were not registered until 1972, and sometimes later. Minor LCVs would have been found in every town and city in the UK. They would have been a popular sight as 'The Post Van', 'The Telephone Van', not to mention all the major utilities, such as gas, water, and electricity. Local Councils also used LCVs in every department, from Repairs to Meals on Wheels and Public Health to the Parks Department. It was not unusual to find the milkman using the Minor van or pick-up. There were a number of companies at this time that would have put specially made bodies on to the basic chassis, thus enabling the LCV to be adapted for almost any use.

    The Minor Commercials were different from the cars, as they consisted of a box-section chassis frame; front end built in unit construction with wing valances, bulkhead, scuttle, windscreen pane, door 'B' post and cab roof, to form the basic vehicle. They were supplied with or without a back panel to the cab, as a chassis/cab without a rear body, in order that you could place you own custom back onto the chassis. They were also available with an all-metal van or pick-up body.

    An additional front seat was available as an extra, on home market models. The van body has large double rear doors, for easy loading. The rear doors have a single-action lock and safety glass windows. The pick-up could be supplied with metal hoops and a waterproof canvas tilt as extra. There was also an option to have a tonneau cover, to place over the load space at the rear. Safety glass was fitted all round, with leather-cloth upholstery and rubber floor matting. A driver's sun visor and exterior driving mirrors, with a one-piece curved windscreen, were all standard. A large parcel tray was also fitted, with provision for heater equipment, again at extra cost. The vehicle also came with a full-width front bumper. These commercials often left the factory in primer, so that various company colours and liveries could be added. One popular back that was added to the chassis/cab was the gown van, used by companies like Burton, John Collier and Kuperstein along with many others in the rag trade. This was an extended body over the cab; it was wider and longer than the standard back, with a tall single rear door, allowing extra height and width in the rear compartment.

    The Minor van and pick-up has also been used as an ice-cream van. There are several different types of 'body conversions'. Nowadays, they are rare vehicles. Mr Seniuk, in Manchester, purchased his new in 1972. It is currently thought to be the only ice-cream Morris van that is still working for a living!

    Builders, plumbers, electricians, television repairer's, garages, bakers, to name a few, would have used the Minor LCV. There are many companies today who see the advantages of having a nicely signwritten vehicle for advertising their business. It also keeps the accountants happy, as all Minor LCV are now tax-exempt. Traders tended to buy their vehicles second hand, from the fleet users of the Minor LCV, who used them for three to five years and then usually sold them off in bulk. These vehicles were used for a few years, but then, after good service, a simple MoT failure would often see them scrapped. Other fleet users of the Minor LCV included the Forestry Commission, Police Forces, MacFisheries, Co-op (all departments), Currys, Burton, Michelin, a number of airlines, as well as various government departments

Most of the utilities (gas, electric and water) are now in the hands of private industry and very little in the way of archives, especially photographs exist. Obtaining photographs has not been easy and has often been a time consuming process. This book has been compiled with help of photographers, friends, friends of friends and many new acquaintances on my search for information and photographs of Minor LCVs. Thankfully, they had the hindsight in the 1960s and 1970s to take photographs of these vehicles, when they were working and in active service.

The Minor LCV can be categorised into the various Series that they were known by.

Series II & Series III (also known as JM Series)

Often referred to as the O type 5cwt (¼ ton) light commercial vehicle van and pick up. Its production period was from May 1953 to September 1962. It was fitted with an 803cc type engine, with an overhead-valve (designated APHM). After October 1956, a larger 948cc engine was available (designated APJM). A four speed gearbox, with a floor mounted gear change lever, 5.00 x 14 cross-ply tyres, (later 5.20 x 14) were fitted, on four-stud pressed steel wheels (4J). A split windscreen was used up to October 1956. After this it was replaced by a single piece. Other changes at this time included a new dashboard, with central speedometer. A hinged passenger seat was added in 1959. The van body capacity was 78cu. ft, with additional 12cu. ft beside the driver. The pick-up, with a fold-down tailgate, also had an optional extra canvas tilt, supported by tubular hoops, with a further option of a tonneau cover, which fitted over the payload area.

Series V

Known as the 6cwt light commercial vehicle van and pick-up. Its production period was from October 1962 to 1971. From 1968, both the van and pick-up were available as 6 and 8cwt It was fitted was overhead-valve, 1,098cc engine (designated 10MA). Specification was generally the same as the Series III LCVs.

Austin Versions known as Series C

This was a 8cwt light commercial vehicle van or pick-up. Its production period was from 1968 to December 1971. The engine fitted was the overhead-valve, 1,098cc. The specification was generally as Series V LCVs. The front kingpins and trunnions were all strengthened to cater for the new rating. 5.60-14 cross ply tyres were fitted on slightly larger four and a half J steel wheels. These vehicles were supplied with the Austin badge and a grille, that differed from the Morris version by 'crinkled' radiator grille slats. There was an Austin motif on the horn push of the steering wheel and it was fitted with plain hubcaps.

Press release, c.1963, for the Morris Minor 6-cwt van, which was available with the new 1,098cc engine from October 1962. The van's silver painted bumper blade can be seen clearly. The cars had a chromed one. *Courtesy: B. Russell*

Cover of the British Motor Corporation (BMC) Passport to Service booklet supplied with every Minor LCV. This particular book refers to the Minor 6-cwt and a number of other BMC vehicles from around 1964. *Photo: D. Leach*

Press Release, c.1968, shows an Austin Minor 6/8-cwt pick-up, which became available from 1968. The Austin bonnet badge and crinkle grille are clearly visible.

An Austin sales advertisement for a Minor, c.1968. Only the commercial vans were badged as Austins. Probably the most reliable van in the world. *Photo: LCV Register/ D. Thomas Collection.*

One
The Minor LCV
(Light Commercial Vehicle)

A Brief History.

After some three years of planning, the LCV range arrived in May 1953. The production of Prototype EX/SMV/163 in December 1949 was the first LCV. It had a side-valve engine, as did the next two prototypes. The first prototype was dismantled in May 1950. The third prototype, EX/SMM/174 (178), which was built on 16 February 1951, did survive and is believed to have been sold as a second-hand blue van with a pre 501 chassis number as OEH 15/500. Prototype EX/SMM/195, from November 1951, is suspected of being the basis of the Cowley Fire Engine, TFC 953, though its chassis bears little resemblance to the usual LCV chassis. This vehicle still exists to this day, and is often seen at MMOC events. The last is the prototype of the rubber winged GPO van. This was EX/FHC13/212, and was built on 29 January 1953 as a 5-cwt quarter ton GPO mail van. It was dismantled in June 1959. Listed below are some of the changes that have taken place in the production life of the Minor LCV. However, this is by no means an exhaustive list.

1953: First van produced as a Series II. The difference from a saloon is that the LCV has a true chassis, telescopic rear shockers, as against levers, a shorter and unchromed front bumper, and the bonnet badge, which was carried throughout it's entire production, which was of the MM type, with a separate chrome strip. Whereas all other OHV Minors had the integral strip and badge. The early LCVs were also supplied with fixed quarter-lights. Late in 1953 these were changed to opening ones. Post office specifications included rubber front wings, with headlights sitting on top; security mesh on the inside of the two rear door windows; a wood partition covered with wire mesh behind the driver; no heater, but an opening on the driver's side windscreen, to cure demisting; and a pre-war roof fitting windscreen wiper.

1954 New dashboard with central speedo, horizontal slatted grill. Rubber wings discontinued.
1956 The first Minor 1000, with 948cc engine and remote box. Improved indicator lever.
1957 Plastic gear knob. Last series II delivered to post office.
1958 Depression added to the gearbox cover, adjacent to the clutch pedal, to increase foot room. Alterations to the door checks to give a wider opening.
1959 Horn button gained independence from indicator lever.
1961 Stronger clutch springs fitted
1962 New 1,098cc engine. Rating lifted from 5cwt to 6cwt. Semaphore indicators at last replaced. Front brake drums increased from 7in to 8in. No hubcaps for GPO or telephone vans.
1963 Fresh air heater introduced. Wiper spindle repositioned. Both front doors given key locks. Zone toughened windscreen fitted.
1964 Large combined flasher and sidelights fitted at front. Wiper blades lengthened.
1965 Slightly redesigned facia. Speedo included a light to denote blocked oil filter. Separate starter button and ignition lock replaced with all-in-one key starter. Wipers and lights now with flick switches.
1968 Vans and pick-ups introduced as Austin badged models. Austin crinkled grill. Austin badge on bonnet and steering wheel. Plain hubcaps. Wheel rims to $4\frac{1}{4}$in, and increasing payload to 8cwt. Steering lock and ignition switch on steering column in some models.

Chassis numbers prefix analysis

From 1953 to 1958 the Nuffield prefix system was used, up to approximately number 149536, although some delivered to GPO were known to have an O-Style prefix, at least to 165449 in 1963. This consisted of three letters and two (later one) digit.

First Letter	Type of vehicle O = LCV, F= Cars
Second Letter	Type of bodywork E= Van, F = Pickup, G = Cab, H = Mail Van, J = Telephone engineers van
Third Letter	Colour A = Black, B = Light grey, C = Dark red, D = Dark blue, E = Mid-green, F = Beige, G = Brown, H = CKD finish, J = Dark grey, K = Light red, L= Light blue, P = Ivory, O = Grey, R = White, S = Mid grey, T = Light green, U = Dark green.
First Number	1=RHD Home, 2=RHD Export, 3=LHD, 4=North America, 5=CKD RHD, 6=CKD LHD.
Second Number	Paint Type 1= Synthetic, 2= Synobel, 3= Cellulose, 4= Metallic, 5= Primer, 6= Cellulose body/Synthetic wings.

Therefore, OEJ1 134747 is as follows: 0= LCV, E= Van, J= Dark green, 1=RHD home.
From approximately chassis number 149537, BMC type prefix introduced as follows:

First Letter	Make of car – M=Morris A=Austin
Second Letter	Type of engine – Always A on minor
Third Letter	Type of body – V=Van, U=Pickup, Q=Chassis/Cab, G=GPO Postal van, E=GPO Engineer's van.
Fourth Figure	Letter or number indicating series of model – 5=LCV, C=Austin LCV, Also from approximately 1968 a suffix letter was introduced to indicate at which factory a particular vehicle was built, *F= Adderley Park. Which appeared after the chassis number.

Therefore, MAV5-255635 is as follows: M= Morris, A Series, V= Van, 5= Series V.

The Minor commercial range were built at Cowley and some were built at Abingdon, there always has been some confusion as to whether the Minor LCV was actually built at the Morris

Commercials' plant at Adderley Park. It is believed from information by ex-BMC employees that production of the van and pick-up moved to the Nuffield Metal Products, Common Lane plant, Washwood Heath, Birmingham, which was to become the Leyland-Daf LDV site. The Minor LCVs were assembled, painted and finally trimmed at the Common Lane site, and then taken by transporters to the Adderley Park plant for the fitting of the engine, transmission and rear axle. Nuffield Metal Products later became Pressed Steel Fisher, and then part of British Leyland.

In the 1960s, Australia had a different numbering system to the UK. Vehicles carried their own compliance plate. I am not exactly sure of the letters and digits, but it is thought to be as follows:

First Letter = Y = Export
Second and Third Letter = series/type of engine (JM= 1098cc and JB= 948cc)
Fourth Letter = ?
Fifth Letter = Type V= Van, U= Pickup, Q= Chassis/Cab, G=GPO Postal van,
First Digit = 3=Series III

The above information is a very patchy, and I would welcome any further details. Most of the commercials in Australia during this period also had an Australian compliance plate, fitted onto the bulkhead, which listed vehicle as 'type 1081', and carried the BMC rosette inside an outline map of Australia. This plate also carried the colour of the vehicle, written and not coded. The colour names were vastly different from those listed in the UK.

Year	Vans	Pick-ups	Chassis/Cab	Total	Chassis Nos.
1953	3374	838	60	4272	501-5214
1954	10992	359	460	15043	5215-21359
1955	?	?	?	18381	21360-3870
1956	?	?	?	13860	38781-52065
1957	?	?	?	16167	52066-68228
1958	?	?	?	15369	68229-84281
1959	?	?	?	15990	84282-100241
1960	13676	3290	211	17177	100242-118555
1961	16614	3819	164	20597	118556-137652
1962	11682	3057	133	14872	137653-153628
1963	12073	2463	20	14556	153629-168223
1964	12506	2596	23	15125	168224-183446
1965	14222	2342	12	16576	183447-202220
1966	11406	2048	8	13462	202221-213679
1967	13652	3288	36	16976	213680-235634
1968	22572	2959	?	25531	235635-258623
1969	20839	3974	?	24183	258624-282895
1970	17210	3083	?	20293	282896-302805
1971	21833	2976	?	28221	320806-325885
1972	2304	463	?	2767	325886-327369

Note:
49767 recorded as last Series II
49801 was recorded as first Minor 1000
Last 948cc possibly 151329 while first 1000 possibly 146403
Austin version starts 236504
8cwt starts 238597
149501 was allocated as being first built at Adderley Park.

Although recorded as the first Austin, the LCV Register knows of an earlier Austin-badged version, with chassis number 235127.

13

The Minor LCV chassis numbers are thought to have started at 501 and ended at 327,369, although an experimental LCV was thought to have been issued chassis number 500 and was sold off as a second-hand blue van. The following chassis numbers were not allocated 49,768-49,800, which is 33, 149,292-159,000, which is 209, giving a total of 242. Which calculates as 327,369 (last number) minus 501 (first number)=326869, minus 242, which gives a grand total built of 326,627.

Below is a table of the known colours used on the Minor LCV. It is not exhaustive; it also includes the paint codes.

Colour Code
Almond Green GN 37
Connaught Green GN 18
Dark Green GN 12
Empire Green GN 22
Everglade Green GN 42
Beige BG 8
Sandy Beige BG 3
Birch Grey GR 3
Clarendon Grey GR 6
Cumulus Grey GR 29
Dove Grey GR 26
Frilford Grey GR 5
Pearl Grey GR 10
Platinum Grey GR 24
Yukon Grey GR 7
Azure Blue BU 27
Persian Blue BU 39
Old English White WT 3
Snowberry White WT 4
Damask Red RD 5
Rose Taupe GR 27
Aluminium

Approximate year that colours were used on ex-works vehicles, please note again this list is not exhaustive:

1953 Platinum Grey, Dark Green, Azure Blue, Beige, Empire Green.
1956 Clarendon Grey, Sandy Beige, Bronze,
1957 As above + Black, Turquoise, Birch Grey – Bronze discontinued.
1959 Frilford Grey, Connaught Green.
1960 Pearl Grey, Yukon Grey.
1961 Dove Grey, Almond Green, Rose Taupe.
1968 Cumulus Grey, Everglade Green, Persian Blue, Damask Red, Snowberry white.
1970 Aqua, Antelope, Teal Blue, Flame Red, Glacier White.

Oldest Surviving Known LCV

In my pursuit for photographs for this publication I wrote to all the MMOC clubs around the world and by pure luck have discovered the oldest known surviving Minor LCV Series II, with a chassis number of 722. Bearing in mind they only started at 501, this makes it very old and pre September 1953, as it has fixed quarter-lights! The van is in Canada and is currently being restored by the owner Nick Harvey, who did not realise it was the oldest known survivor, until I told him! The full details are OEH45 722, last registered in 1963 as T4734. O = LCV, E = Van, H = CKD Finish (primer?), 4 = North America, 5 = Primer, 722 = pre September 1953.

All in all, this makes it the oldest surviving Minor Series II LCV the 221st built, easily overtaking OHP 900, OEH15 2697, which has held title for a number of years. I am also informed that since 722 was located, while the chassis was number being verified, Pete Hanby, who runs the Series II register, had discovered an earlier LCV than OHP 900, OEH15 2697; this being OEH55 1200. OEH = as above, 5 = CKD RHD, 5 = Primer, 1200 = 699th LCV registered. Mark Malloney owns this vehicle and it is currently under restoration in Auckland, New Zealand.

Nick sent the registration document and a rubbing from the chassis plate confirming the vehicles chassis number. I was still able to inform Nick that his van was the oldest in the World. 722 is in very good shape, body wise, and still retains its original 803cc engine. This very rare preSeptember'53 Series II was originally used by a plumbing firm in Canada that went under the name of Roto-Rooter, this was painted on the side panels. The logo has been reproduced from an advert in the Canadian Yellow Pages. Under the logo, on the side panels, is a ditty, which starts and ends with a musical note and reads 'And away go your troubles down the drain'! They used the vehicle from 1953-1964, where it was parked under a bush. The vehicle remained there until 1975, when the previous keepers discovered it and purchased it. The vehicle then remained in a garage, gathering dust, until Nick purchased '722' in 1998.

The previous keepers started restoration in 1992, but never seemed to get very far, as can be seen from the photos. However, I am assured that will not be the case for long and that the vehicle will shortly be returned to its former glory, which is only what a vehicle, this rare deserves. We all look forward to the day when the vehicle is restored, hopefully by 2003, the fiftieth anniversary year of the production of the Minor LCV.

The Newest Van in the World?

John Wass of Hull, informs me of a new van in Hull. Hull School of Technology purchased a new LCV in the 1960-70s and was used as a rolling chassis by the School of Technology. A 'new' body and engine were later acquired, but the vehicle was never registered, or for that matter even assembled until a few years ago. This un-registered new Minor van has since been passed to the local transport museum, where it has been painted in the blue and yellow livery of the Hull City Council engineering department. The new van is now officially a member of the council's fleet, but still remains un-registered and is likely to stay that way.

The oldest known surviving LCV in the world. The van is currently under restoration in Manitoba, Canada. It started life as a plumber's van. *Photo: N. Harvey.*

The chassis plate from '722' indicates that this was the 221st LCV off the production line in 1953. The van also has its original 803cc engine. *Photos: N. Harvey.*

The second oldest known surviving LCV in the world. It has a chassis number of 1200. It is currently under restoration in Australia. It was discovered at the same time as '722'. *Photo: M. Malloney.*

LCVs (Series II) being assembled in Australia c.1956. The vans on ground level are in an unfinished state, whereas the pick-ups on the inspection ramp seem to be completed. All the LCVs have chromed bumpers. *Photo: P.M. Photography.*

The Australian compliance plate (equivalent to the UK chassis plate). This chassis numbering system is unique to Australia. The colours are also written out in full. Few Australian colours are listed in any Austin/Morris colour charts. *Photo: B. Sharman.*

17

587 KYC, a 5-cwt Morris Pick-up, c.1960. On the left is Sam Cook with his new pick-up, complete with canvas tilt. On the right is his good friend George Lawrence. The vehicle has already been modified with A35 sidelights on the tops of the wings, and Mini indicators.

587 KYC, c.1975, looking the worse for wear! It now carries the livery of the garage owned by Sam, where the pick-up was used as a garage runabout. Somehow or other it ended up in a ditch, with Sam at the wheel. Near the rear lights a bicycle can be seen, which was being ridden by George, who had accepted a lift from Sam. The bike survived the incident and in the end they both ended up riding the bike home! The next day the pick-up was recovered.

587 KYC, after restoration and repairs in 1999 The pick-up has returned to its original livery and still looks good! *Photos: S. Cook*

Postcard of Coventry Road, c.1965. The hardware shop seems to stock everything for the road user, but it can still find time to service a Morris delivery van, complete with period roof rack. *Courtesy: L. Fleming*

Just purchased! A brand new Austin Minor van, c.1968, which cost around £420, from Alsops Garage, of Newport. It has plain hubcaps an Austin grille and bonnet badge.

A typical BMC garage sales forecourt scene, c.1968, at Alsops Garage, Newport. The commercials in the photo include Mini pick-ups and vans, Land Rovers, an Austin Gypsy, an A35 van, and even a Ford Transit. Towards the centre of the photo are four unregistered Minor vans, priced at £435, which would have made them the more expensive 8-cwt version. *Photos: J.B. Williams*

A working LCV, *c*.1980, at Victoria Tyre Depot, in Holywell. On the right of the picture, in the shade, is a Mini pick-up, which is fast becoming a collectable vehicle. *Photo: T. Bourne.*

Bill Wilkinson poses with his Austin 8-cwt van, which he had just finished restoring, complete with replica livery, in 1991. The van, with its $4\frac{1}{2}$ Js, also includes an illuminated Bibendum, the Michelin man, on the cab roof. *Photo: B. Wilkinson.*

A chassis cab was an option available from the works, but not as basic as this! This is a completely stripped-down van. The box section chassis can be seen, unlike in the car. Everything, including the cab bolts onto a true chassis. This vehicle, FPK 477J, belongs to Peter Willmet. It was restored to look like the van below. *Photo: P. Willmet.*

Maidstone & District bus maintenance van, c.1967. The livery was all over mid-Brunswick green with light green side panels, and Maidstone & District embossed in gold. They had an Austin and a Morris Minor van, which were used for attending breakdowns. *Photo: Maidstone & District/P. Willmet Collection*

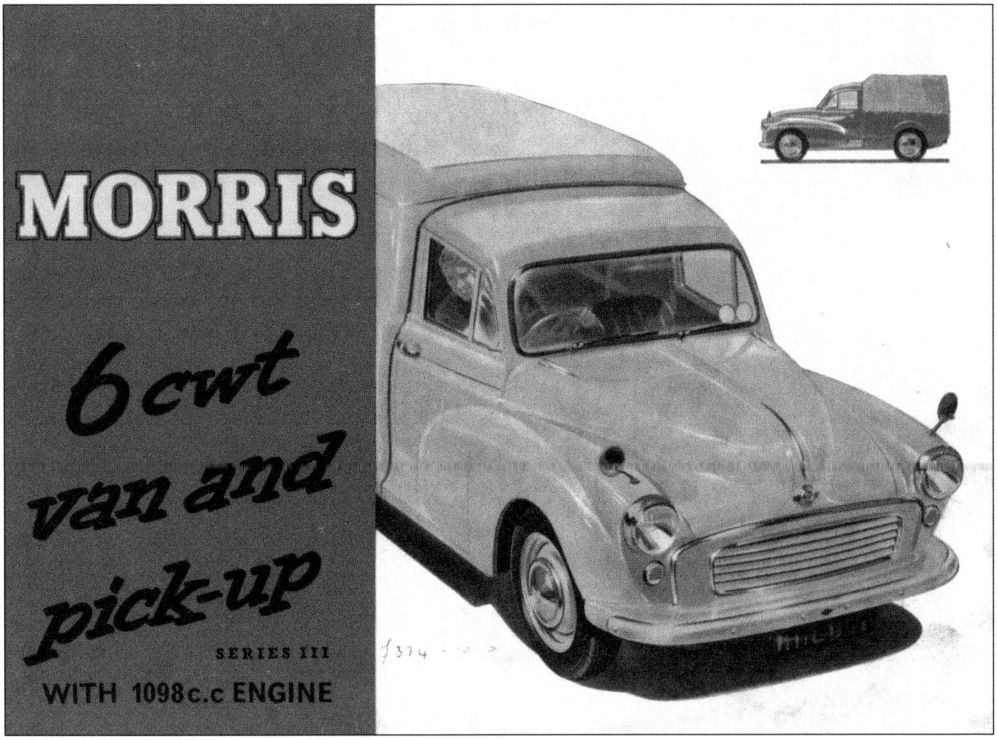
MORRIS
6 cwt
van and
pick-up
SERIES III
WITH 1098c.c ENGINE

The cover of a full colour sales brochure, *c.*1963, for the Series III. It boasted an 'increased payload brings better business'. Details of the LCV's specifications and dimensions were shown. It also contained drawings of the tradesmen who would have used the Minor. *Courtesy: G. Crew.*

SUPPLY & CONSTRUCTION
ACE
READY TO USE
BMD
TILE CO. LTD

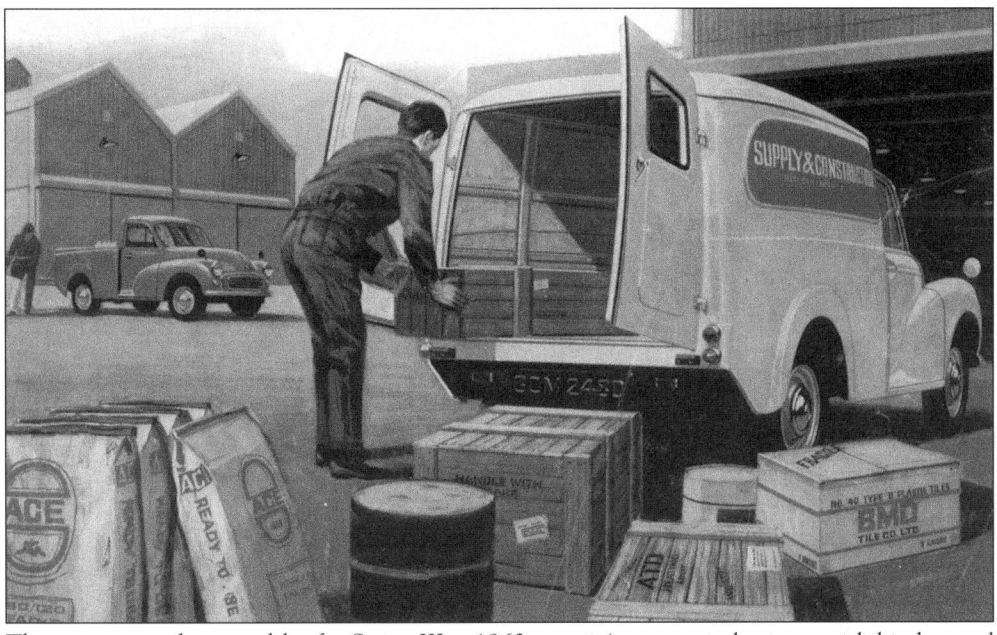

This two page sales pamphlet for Series III, *c.*1963, says it 'puts you in business with big-hearted low transport cost'. It was printed in blue, black and white only. *Courtesy: R. Harvey.*

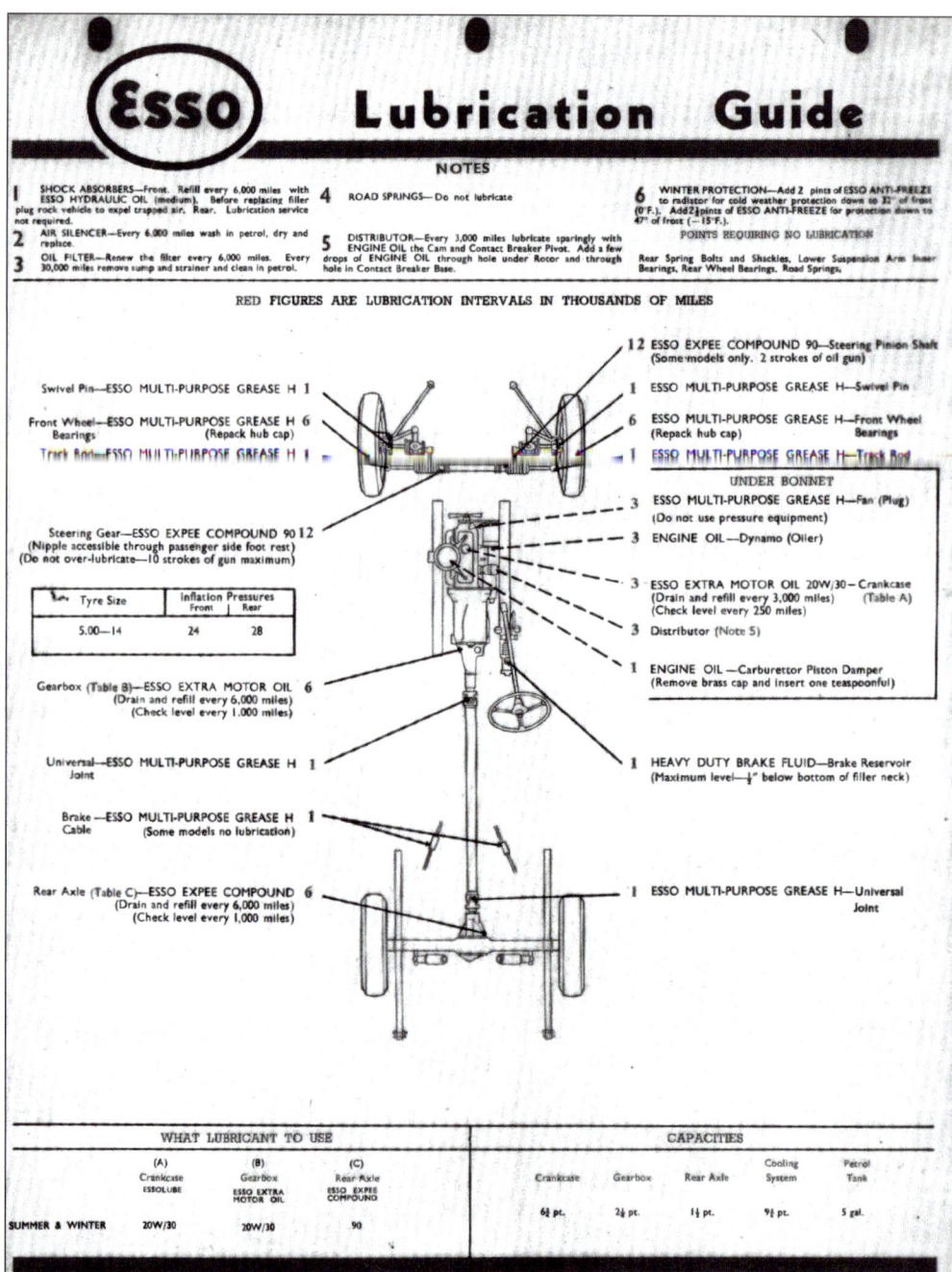

An Esso lubrication chart for the quarter-ton van, c.1958. Courtesy: B. Wilkinson.

An ex–post office engineer's van, JLT 675D, c.1966. Here it was being used by MacDale Garage in Broughty Ferry, near Dundee, in 1978. *Photo: R. Doig.*

A new looking Empsons Teas Minor van, c.1960, in Birmingham. The shine on the van is so deep the houses opposite are reflected in the side panels. *Photo: J. Murphy.*

Barry Allen bought this van from new in 1970 in Liverpool. He originally purchased it for his wife's business. It has rear windows in the side panels, and was fitted, from new, with rear chrome bumpers from a Minor Traveller. It is shown in 1979. *Photo: B. Allen.*

Here is a Minor pick-up, pictured in 1979. It was bought in 1977 to replace an earlier model first registered in 1969, as a recovery vehicle for use in Mr Estlea's motorcycle business. *Photo: D. Estlea.*

A quarter-ton Series II pick-up is a rare sight these days. This vehicle dates from around 1955, although the photo was taken at Mullion Cove in Cornwall around 1979.

An ex-Royal Mail van, c.1971, at Polpero in Cornwall, eight years later. *Photos: M. Forster.*

Rolls Royce dealers P.J. Evans in Birmingham used this Minor van c.1972, GFD 23K, for ferrying parts to various branches around Birmingham. Here it is shown, during restoration, in 1998 at the gates to the former premises of PJ Evans dealership. *Photo: B. Lee.*

Ex-PO engineer's van WFS 500K, c.1972, enjoying a new life as a Hoover service van, is seen here on 9 May 1979 in St Andrews. Note the unusual roof-mounted spare wheel and off-set aerial. *Photo: R. Doig.*

Colin Yarwood used this former GPO van, c.1967, to transport stock cars to and from the circuit in 1974. The car on the trailer is a Ford Anglia 105E. *Photo: C. Yarwood.*

The GPO still used Series II up to 1958. TUV 399, c.1958, is looking rather the worse for wear in 1961! The van was probably a write-off. *Photo: M. Foster.*

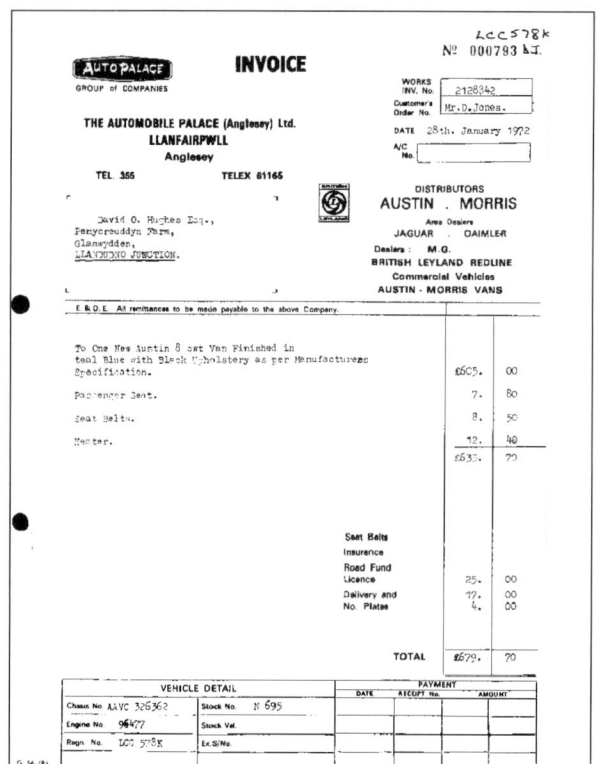

Here is the original bill of sale, from 28 January 1972, for one of the last Minor LCVs produced, an Austin van LCC 578K, sold to a Mr D Hughes in North Wales. *Photo: B. Rhea Collection.*

Austin van LCC 578K in near original condition, in 1999, now owned by Brian Rhea. *Photo: B. Rhea.*

Fish By Royal Appointment, 1961. This Mac Fisheries van worked from the Windsor shop. It was the only branch that was able to carry the royal crest. Every shop had at least one van. They would have purchased in excess of 500 vans. At the end of the day drivers used to park the vans on ramps and hose them down inside, in an effort to remove the smell of fish! *Photo: B. Russell.*

This Series II ex-GPO engineer's van NLW 918 c.1954, was being used as a shed on a farm in north Wales in 1982, when Martin Forster found it. He then painted it blue. BT bought it and apprentices restored its original GPO livery. Note the rubber wings, which were only fitted to GPO vehicles, and in particular the dents in the nearside wing. *Photo: M. Forster.*

Sparks of Streatham was a dealership for British Leyland's new van. Here it is shown on its first day of service in 1968. This van has since been restored, and, with its current owner, has appeared on TV. *Photo: J. Murphy.*

Here is an application form for the BMC Drivers' Club, which was formed in 1955. The club catered for all Austin and Morris commercial vehicles. It offered low-rate insurance, safe driving awards and a club journal, called 'Sidelights'.
Courtesy: G. Crew.

Two
GPO, PO and Kingston Telephone Vans

The GPO was by far the largest fleet user of the Minor LCV. During the autumn of 1953 it took delivery of the then new Minor LCV quarter-ton vans, which were powered by the 803cc engine. The Post Office Minors can be separated into two groups, Royal Mail and Post Office Telephones. They differed from the normal production Minor LCVs. The Mail vans were referred to by their load-space capacity, that is cubic feet, in the case of the Minor, it was fifty cubic feet, and they were recorded as 50cfs. Whereas the PO telephones were known by their laden weight (5 or 6 cwt). The Minor LCV range normally mirrored the production changes of the Minor saloon. The first Post Office Minors had rubber front wings, which were fitted until 1955. These vehicles are now very rare, but are occasionally seen on the rally field (Peter Hanby has a fine example, PGO 296.)

All Minor commercials had rubber matting was on the floor; later versions had rubber linked mats. The front bumpers were painted silver, not chromed, and rear bumpers were two small rubber blocks, located under the rear doors. In addition to the rubber wings, that had headlamps mounted on the top, the driver's half of the split-screen opened outwards and wipers were fitted above the screen. The plywood fascia on the dashboard was changed with the demise of the rubber wings. There was no heater. A blanking plate was fitted to the cylinder head, in place of the heater tap, to avoid it getting misted up in bad weather.

One of the earliest production models was registered NLW 918. It has now been restored and is in the BT Museum, in London.

The rubber wings, which had perched headlights, were only in production for a year or so. They were replaced by standard steel wings and headlamps. These vans carried the normal Morris badge and bonnet flash, which was carried over from the MM days. They remained on the commercials until the end of production in 1971. They were originally supplied with hubcaps, although this was changed to keep cost down in 1962.

The rubber wings that were fitted to the Series IIs GPO vans, always seems to have the same dent in the nearside wing. All the existing Series IIs are effected in the same way. The damage looks as if someone has sat on the wing. The only sensible explanation for this has come from Pete Hanby of the Series II register. He believes that the rubber wings were stored together, with the nearside bent inwards and the offside outwards, thus causing the dent in the same place on all the Series IIs.

The rear door windows had wire mesh fitted as a security feature and a wood framed partition, covered with mesh, was placed behind the driver, which separated the van cab from the rear. A further safety on the Royal Mail vans was that the rear doors could be locked from the cab, by a special lever operating a locking drop bar. Post Office Telephone vans had a special hasp, until 1969, on the rear doors, to lock them with a padlock. They also had a roof ladder, rack and ladder. Yale locks were fitted on both front doors of mail vans from 1958, which provided extra security, so that the postman could leave the van to empty the post-box without having to lock the doors, as long as he remembered to take the keys to get back in.

The early vans were fitted with trafficators, until 1958, when they were replaced with indicators. The early ones were positioned on the van back roof, in the middle of either side and acquired the names 'bats' or 'pigs ears'. In time, they were put in the rear lamp cluster, in line with the rest of the LCV range.

Some early PO vans, up to late 1960s, have a 2in 'M' or an 'E' pressed into the floor moulding, just in front of the rear wheel arch. This represented the different pressings for different cargo floors for the various PO and GPO vans, due to the location of the captive nuts.

During 1962 both mail and telephone vans were fitted with fastenings on their sides, for carrying placards, informing 'how little it cost to telephone Australia' or 'When to post Christmas cards in time to reach Canada' and other such like messages. The fastenings were higher on the side panels on the PO Telephones van than on the Royal Mail van, due to the curved lettering of 'Royal Mail'.

Unlike the rest of the commercial range, the Post Office Minor LCVs never used the 948cc engine. In 1962 the payload capacity was increased to 6-cwt and the much-needed larger sized 1,098cc engine was added for GPO use. These vehicles became known as Series V and from 1968 there was the option of an increased payload to 8cwt, as well as retaining the 6cwt. They were also badged as Austin and were known as Series C, the post office decided not to use them. Although the Series V had arrived, along with a new chassis numbering system, i.e. MAV 5 xxxxx, the Post Office still retained OHC and OJE until 1967.

The livery carried by PO Telephone vans was all-over mid-bronze green. There was a crown, but no EIIR initials, on the side panels with 'POST OFFICE TELEPHONES' and 'TELEPHONE MANAGER'. The area in front of the rear wheel arch and behind the front doors was white. At the bottom of this panel is the vehicle fleet number. Also on the front and rear arches are the tyre pressures, again in white. Towards the end of production these vehicles were supplied in all over yellow livery, with all writing in mid-green. The only difference to the earlier livery was that the vehicle fleet number appeared on the van back roof, over the cab and over the back doors. Some of the later Welsh vans had a large telephone handset on the side instead of the words 'PO TELEPHONES'. PO vans up to 1968, were painted cream inside. Then they were painted the same colour as the outside of the van.

Royal Mail vans carried an all over red livery, with an arched 'ROYAL MAIL' written on the side panel, with a crown and the initials 'EIIR' (Elizabeth II Regina) beneath it. Up to 1968 they were painted cream inside, after this date they took on the colour of the outside, like the PO vans. Also on the front and rear arches were the tyre pressures, in white. These vehicles also had a large roundel on the near-side door. It was a copy of a postmark, round with two wavy lines, and had 'POST OFFICE' along the top and the 'DISTRICT' to which the vehicle was allocated along the bottom, with the number in the middle. From 1969 Welsh vans had 'POST BRENHINOL' on the side, as well as the English translation. This was due to the investiture of Prince Charles as Prince of Wales. Scottish vans are painted in the same way, but the crown has no initials and was slightly wider. This is because when Queen Elizabeth II came to the throne in 1952, the St Edward crown was adopted, but she was not the second Elizabeth in Scotland and the St Edward crown did not apply north of the border. So, the Queen's crown of Scotland was used. The same problem occurred with the K6 telephone kiosk. They were manufactured with a slot to insert the Royal Crown depending where the kiosk was placed.

Towards the end of the 1960s, the liveries of both the PO and mail vans changed. Mail vans became a brighter shade of red, while the PO vans changed dramatically from a drab olive green to a bright golden yellow. Earlier vehicles got this livery when they were overhauled and repainted. The prototype for this livery was GLK 57C.

From 1969 Postal Engineering vans were supplied in the Royal Mail livery of allover red with a 6in white band painted around the waist of the van back. Some also had a white stripe wide over the bonnet and the cab roof. Written on the white stripe at the rear was 'Postal Engineering' in black and a gold crown. These vehicles carried the roof ladder rack. The drivers checked the Post Office scales and also emptied the coins from the telephone kiosks, which, strangely enough, is still done by the Post Office today despite its separation from British Telecom.

During the 1960s another engineering van livery was introduced. It is known to have worked in the Leicester area, but whether it was adopted nationally is unclear. The Postal Engineering vans, with the white stripe around the waist and over the cab, were used on survey duties only. But green PO vans, with a six inch red stripe painted around the waistline, but not over the cab, were used on duties such as repairing the scales at post offices, lights and stamp machines. One Minor is known to have made it into the BT livery of all-over yellow with the Blue early BT logo on its side panels, the registration number was CKN 322K.

The vast majority of GPO vehicles, like other government departments, carried London registration marks, issued by the London County Council. In autumn 1953 the first deliveries of production GPO Minor LCVs were allocated registrations in the NXO and NLW series. Both GPO block registrations were shared with many other types of GPO vehicles. 1963 saw the issue of xxx GPO, and again in 1965 GPO xxxC. Not surprisingly, these were reserved for GPO vehicles. (PO was a Guildford registration mark.) Many of these were reregistered upon sale, but several managed to slip through the net. 331/412 GPO, GPO 84/106C are known to have survived among others. Around one hundred and twenty of the batch of GPO block registrations allocated were voided in favour of Guernsey, Jersey and Isle of Man local registrations. 100 BLE would have been the Bahamas. Block registrations for GPO Minors continued until October 1969, when the Post Office Corporation was formed. From this point, all vehicles were known as PO vehicles and carried local authority registrations.

As well as the differences mentioned, Royal Mail and Post Office Telephones would have had numerous features unique to the GPO and PO. Colin Ellis has helped me to compile a list of the typical features that apply to late vans, from 1970.

1970 Royal Mail Van

These vehicles would have had one seat, metal door panels and a switch, mounted under the dashboard, that had GPO written on it, which controlled the rear interior light. A fire extinguisher was fitted to the passengerside quarter-panel. The quarter light catches were sawn off and window stops were fitted, for security. The toolbox was located behind the driver's seat. A starting handle and first aid box were mounted on the wood framed wire mesh partition, behind the driver's seat, with the document basket and T-bar for the toolbox.

They had mud flaps, front and rear, and a chrome pull handle on the offside rear door. Rear doors have hooks on the outside panel and there were rubber buffers on the bonnet corners. The vehicle was also equipped with a towing bracket on the rear off-side chassis, u-bolts on the front arm, to stop the jack from slipping, GPO wing mirrors and double-sided wheel nuts. Under the bonnet there were electric washers and a battery tray.

1970 Post Office Telephone Van

These vehicles had two seats, with metal door panels and a switch mounted under the dashboard, that had GPO written on it, which controlled the rear interior light. A fire extinguisher was fitted to the passengerside quarter-panel. The quarter light catches were sawn off. A starting handle and first aid box were mounted on the woodframed mesh partition behind the front seats. The spare wheel was also mounted centrally on the partition in the cargo hold with the document basket The toolbox, which had a hasp and padlock, and was located behind the passenger seat.

They had mud flaps, front and rear, and poster board clips on both sides, higher than the Royal Mail vans. Rear doors had hooks on the outside panel and rubber buffers on the bonnet corners. The vehicle was also equipped with a towing bracket at the rear, u-bolts on the front arm, to stop the jack from slipping, GPO black plastic wing mirrors and double-sided wheel nuts. There were electric washers and a battery tray under the bonnet. There was an aluminium roof ladder rack, with aluminium storage bins located in the rear on both sides of the van, plus a chain to secure the interior ladder in place.

The GPO and PO purchased in excess of 50,000 Minor LCVs. They were the largest fleet user of the Minor van, although some of these vehicles were even registered on 'M' (1973) plates.They also used it for the TV Licence Investigation department, which was renamed 'Subscription Services Ltd' in 1986. They were first used in Scotland. The livery was thought to be orange. Then they had a light-blue with the crown on the side panel, with 'EIIR' and 'TELEVISION LICENCE INVESTIGATION' below. It is not known how many vehicles were converted, but they were thought to be ex-mail vans. They were often fitted with a new engine, but did not have any detecting equipment. The only registration known is UBO 882J, which would have carried serial number 273450 with the Royal Mail. It would have been changed to the 167xxxx around 1975 or 1976, on taking up new duties.

Minor LCVs were also used as driver under instruction vehicles by the GPO and PO. They normally had two Perspex opening vents, one-foot square, mounted in the van-back roof, with a seat and light in the van back for the examiner! Some were dual-controlled, like XKG 620K. They had a large 'L' plate in a flap in the front grill. The rear L-plate, was mounted on the rear offside door, on a special bracket. Both read 'DRIVER UNDER INSTRUCTION' as well as having a large 'L'. The following vehicles carried all the liveries of the PO and GPO: Royal Mail red, XKG 620K, recently restored and owned by John Bowen, PO Telephones yellow, GLK 95C, and GPO Engineering red and white, WLA 555G. It is not known how many vehicles were converted to dual control.

In 1969 the Isle of Man, Jersey and Guernsey Post Offices became independent of the British Post Office, but they continued to use the Minor LCV. The Guernsey Post Office's livery was mid-blue and sign written 'Guernsey Post Office'. The Jersey Post Office livery was red for the bottom half of the van and yellow on top with 'JERSEY POSTAL SERVICES' arched on the side panels. Both the Northern Ireland and Isle of Man GPO vans had the same liveries as the mainland. The Republic of Ireland used the Minor LCVs and had a dark-green livery.

The Swedish and the Danish, and possibly the Norwegian, post offices used the Minor LCV The Danes used both the standard LCV & the DOMI van. Turkey and Iran also had Minor LCVs mail vans. Both countries opted for GPO specifications.

In the UK Hull and Portsmouth local councils ran their own telephone companies. The rest of the network came under Post Office (GPO) control in 1911. The Portsmouth system only remaining in private hands until 1913, when it was sold to the Post Office. Hull on the other hand remained as the only locallyrun telephone system. It still survives to this day, although its name has now changed to Kingston Communications and has recently been floated on the stock market. Like the GPO, Hull Telephone Corporation used the Minor LCV from the early Series II until the last vans rolled off he production line. The Hull Telephone Corporation are known to have had three liveries. The first was dark grey, with black front and rear wheel-arches. It also included the raised lower panel on the front doors, as well as the front wings. The second livery was the same, but the top half of the van from the waistline up was a very light grey, including the bonnet. Both liveries also included a large roundel on the side panels, with either a 200 Series (Bakelite) telephone on the first livery or a 706 telephone (the first plastic telephone produced) on the second, in the middle of the roundel. It also has 'KINGSTON-UPON-HULL' on the top half of the circle, with 'TELEPHONES' on the lower. The third and final livery was mustard yellow, with 'TELEPHONES' on the side panel, with the 'three crowns of Hull'. All three liveries had written on both front doors 'Telephone Manager, Kingston-upon-Hull Corporation', just below the handle. Some of the vehicles also carried a wooden ladder rack, and some of the early Series II vans also had the 'pigs ears' indicators.

The Crown is used on all mail vans in service throughout the country. However, in Scotland the present Queen is not recognised as the second Queen Elizabeth. Therefore, a larger crown was used without EIIR under it.

The Crown used on all Post Office telephone vans. *Photos: R. Harvey*

Post Office Royal Mail van, *c.*1969, spotted on 31 August 1978, in Brecon Town. The words Royal Mail are written in black, whereas 'POST BENHINOL', the Welsh translation, is in white on the side panels. Note the lack of hubcaps, the white GPO wing mirrors, and the clips for attaching advertising boards. *Photo: Ken Porter.*

This GPO engineer's van, *c.*1968, with the six inch white stripe, shown here in Glasgow in 1974. This van has black GPO mirrors and the small gold crown. 'Postal Engineering' can be seen on the white stripe between the passenger door and rear wheel arch. *Photo: R. Doig.*

A 1950s PO telephone engineer's van, outside a K6 telephone box near Rugby. PGO 296, was one of the first LCVs to be restored by Pete Hanby, who now runs the Series II LCV Register. The dent in the rubber wing was caused by storing a pair of wings together. *Photo: P. Hanby.*

Royal Mail van XKG 620K Driving Instructor was fitted with dual controls, in Cardiff in 1972. The two roof mounted opening vents allow both air and light into the rear, as this is where the examiner would have sat. There was also a specially adapted hand light fitment located in the rear. John Bowen has restored this van to its original condition. *Photo: M. Street.*

Royal Mail van VUF 439K c.1972 (serial number 1040525), is seen at Hounslow Post Office around 1974. The Crown and EIIR, advertising increased postal charges, stand out. Worth noting are the mudflaps, which were a standard GPO fitment, front and rear. *Photo: K Porter.*

The absence of 'EIIR' from the crown suggests that RVS 482J, c.1971 (serial number 273615), is a Scottish Royal Mail van. This angle shows up the post office 'roundel', similar to a postmark on a letter, with the district that the van was allocated to as well as the Yale lock. Photographed at Calton Road, Edinburgh in 1974. *Photo: R Doig.*

This Royal Mail van, UBO 882J, c.1971, shown here in 1975 in Park Street, Cardiff, was converted for use by the TV and Radio Licence Investigation department. The clips for holding the advertisement board and Yale locks, confirm it to be a mail van. The van carries a light-blue livery. In the background is the old Cardiff Arms Park, which has now been replaced by the Millennium Stadium, which hosted the 1999 Rugby World Cup. *Photo: M. Street.*

Here is another mail van turned TV detector, with its striking light-orange livery. It was spotted on 18 March 1970 at Willison Street, Dundee. Unusually, the crown on this 1965 model is placed between the driver's door and the wheel arch. *Photo: R. Doig.*

Post Office Telephone engineers van SLO 944F, c.1968, was seen in Southend-on-Sea. 'POST OFFICE TELEPHONES' and the allocated area are marked between the driver's door and rear wheel arch. The tyre pressures are also indicated in white over the wheel arches. *Photo: POMM/C Stevens Collection.*

Two mail vans, PFG 977J and NSP 970H, parked outside Johnston's Garage, Market Street, St Andrews in Scotland, in the mid-1970s. The van on the right is fitted with town and country tyres at the rear. The rough tread is clearly visible, as are the Scottish crowns. *Photo: R. Doig.*

Hiding in the rushes is NYV 102E, *c.*1967, a mail van on its rural round at Sutton Spring Wood, Calow. It had covered 76,500 miles when it was spotted here on 17 October 1973. It was to go and record over 80,000 miles before being sold off. The winter sun is shining on the front of the van, which highlights the screen behind the driver's seat. *Photo: D.J. Foster.*

Newport, Isle of Wight sorting office *c.*1971. The van CBY 393D dates from 1966. The number plate is mounted on the cab roof, in order not to damage it, and to provide extra ground clearance. This usually indicated a rural round. In the background is a Morris J4. *Photo: A. Browne*

St Helier's transport manager, Mike Green, poses with a brand new mail van CBY 774D c.1966, at the PO garage. The van would have been re-registered as J 17515, but this photograph confirms that original allocation of registrations for PO vans were carried from the supplier and only changed upon arrival. On the right J 15487 would have been EYV 605C and J 11214 would have originally been allocated GYF 270C. *Photo: M. Green.*

Leaving Ivy Arch Road sorting office in Worthing in 1970 is mail van JPO 794C, c.1965, which had been re-registered from the GPO xxxC series after the General Post Office was abolished in 1969. The Minor van in the background is full of mail and parcels. *Photo: M. Skillen*

One of the earliest Series II PO Telephone vans NXO 611, complete with roof ladder, rack and rubber wings, going about its duty c.1955 *Photo: B. Jenkins Collection.*

This was the only Minor to make it into British Telecom livery. It is seen here in 1981. *Photo: M. Skillen*

PO Telephone van ELH 258C (serial Number 65 300 6451) dates from 1965. It is seen here on Christmas Eve, 1973, in New Wynd, Montrose. It was being used temporarily as a mail van.

Two important features worth noting on this new Post Office Telephones' van MYJ 782J (serial number 71 0089), in 1970, are the change of livery from green to yellow and the use of a local Dundee registration number, rather than GPO block registration numbers. *Photos: R. Doig.*

For Sale: three Minor LCVs (two PO Telephones and a Royal Mail), one careful owner, the GPO! What a bargain at £350 each from a garage in Derbyshire in 1979. *Photo: R. Doig*

Frank Rice-Oxley bought this out of service Telephone van WDP 398J (serial number U257476) for his daughter, pictured to his right, in 1980. It was used in and around the Reading and Slough area. *Photo: F. Rice-Oxley.*

The wing mirrors, 'bats ears' indicators and most of the chrome work ended up being yellow when the painter got carried away on this repaint! Post Office Telephone van c.1960. This photograph was taken in 1969 in Birkenhead. 37 GPO has yet to be re-registered. *Photo: B. Jenkins.*

Brand-new WEL 23J PO Telephones' van (serial number U259226) delivered in the bright yellow livery to Bournemouth in 1971. *Photo: B. Jenkins.*

Driver instructor dual control PO Telephones' van, c.1965, shown here 18 August 1973 at Ashford workshops. *Photo: K. Porter.*

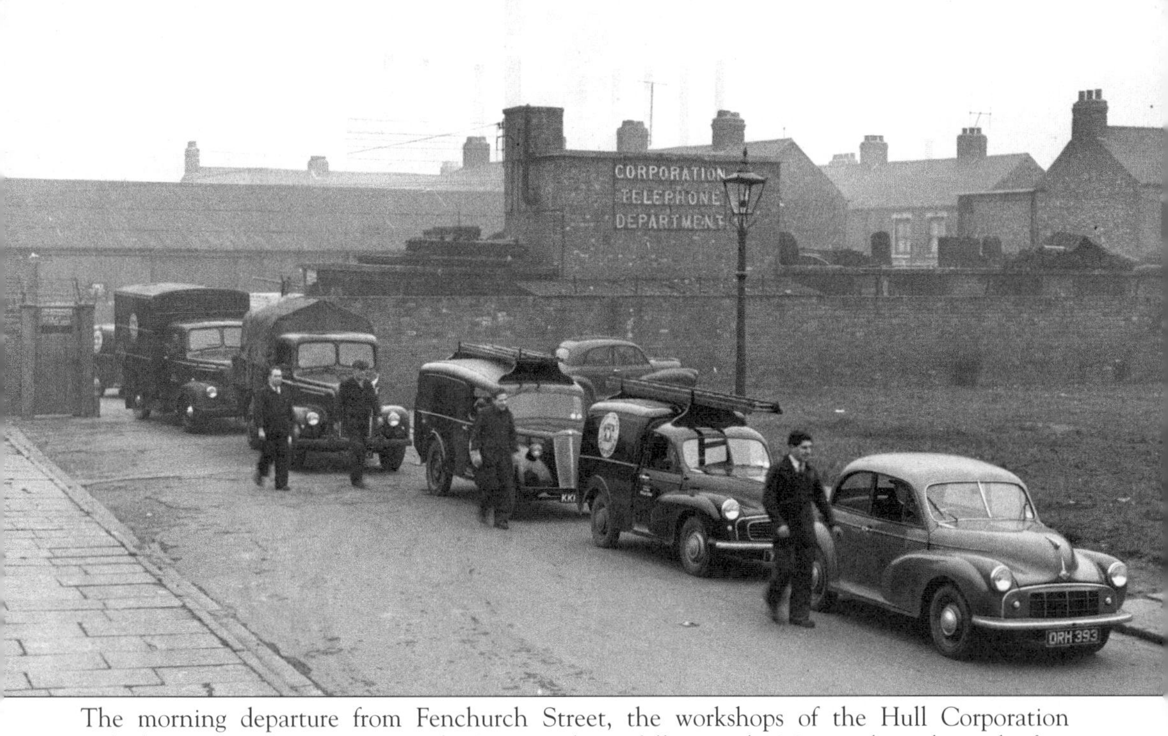

The morning departure from Fenchurch Street, the workshops of the Hull Corporation Telephone Department in 1956. The Series II that is following the Minor saloon shows the first livery that this company carried. It is also fitted with a roof ladder rack, which is totally different from that of the GPO specifications. *Photo: Kingston Communications*

A Kingston-upon-Hull Telephones' van displaying the company's second livery on the quayside at the docks in 1965. *Photo: Kingston-upon-Hull City Council.*

The third and final livery of Kingston-upon-Hull Telephones that was carried by Minor vans is shown on an A Morris 6-cwt Series V van, in the service department, 1970. *Photo: Kingston Communications.*

Three

Adapted Minor LCVs and Unique Features

The Minor LCV has generally been accepted around the world in its standard form of van and pick-up variants, but there have been numerous vehicles converted or equipped with unique features for a particular need or trade. Police vehicles and roadside assistance vehicles also fall into this category, however they are dealt with in a separate chapter.

Conversions ranged from the fitting of aerials to complete body conversions. While researching the D.O.M.I. (Dansk Oversoisk Motor Industrie) van, another rare conversion was discovered. This vehicle was part of a fleet specially converted for the Swedish Post Office. The conversion took place at a factory in Vastervik, Griparosser, in Sweden. Not unlike the Danish van, it is level across the roofline, but it is about ten inches higher than a standard van, and it continues over the cab as well. Its overall length is extended with square wheel arches. A 1954 example is known to exist and is currently under restoration in Lund, Sweden. Another Swedish PO van has been found, but its current owner is still being traced in Sweden. I am indebted to Urban Mattson in Akersberga, Sweden, for his efforts in tracing the second vehicle. These vehicles were certainly in use in Västeråsand and around Ronneby, until 1959. The livery was yellow with black wings and grille and the blue 'Royal postal Horns' on the side panels.

Simon Marsbøll of the NMMK tracked down a D.O.M.I. van. These vehicles we now know were exported as chassis-cabs to Denmark and D.O.M.I. then carried out the conversion on the behalf of the Danish Post Office. This conversion was carried out at the factory in Glostrup, outside Copenhagen. Apart from the Danish-bodied Minor vans, I understand that the Minor saloons were also assembled here from CKD (Completely Knocked Down) kits.

The D.O.M.I. conversion entailed fitting a coach built rear to an imported chassis/cab. The width and the length of the rear section was extended. But unlike a standard LCV it was the same height as the cab roof. It also had a one-piece rear door and a larger rear window, that must have improved visibility. The rear number plate was centrally recessed and the van had rear painted bumpers, in the body colour, not unlike the traveller ones. It also had cab-roof mounted indicators, just behind the front doors. The vehicle also had a unique trim of cream and black check on all three door panels, inside the glove boxes, down to and including the parcel shelf, the rear door appearing to be padded, with the check trim coming into the rear of the vehicle and also covering some of the rear roof. It may have been for noise prevention or insulation, due to the cooler climate. The livery was all over grey, unlike most Danish Post Office (PO) vans, which were mustard-yellow. It still retained its fleet number at the bottom of the panel behind the front doors and forward of the rear wheel arch. Town and country rear tyres were the norm on these PO vans. All three of these vehicles are extremely rare and are thought to be the only survivors of their particular types.

Another conversion, discovered by Dave Thomas, was completed by Stewart & Arden, of Wembley, around 1953-1954, for Pall Mall cigarettes. There are some unique features of this fleet worth noting, particularly the 5-ft cigarette on the roof! Does any body know if this was illuminated? It has chrome wing mirrors, chrome bumpers, with over-riders, and a chrome 'cheese-grater' grille. It is also worth noting that indicators have been added underneath the headlamps and interestingly a fog/spot lamp to the front near-side. The rear wheel arches seem to be very Traveller-like. Lastly, the whitewall tyres and paint scheme, thought to be red and white, gives the vehicle an American feel. Stuart & Arden were also involved in supplying 'WRVS' vans, in some format.

Brian Lee, from the Midlands, informs me that during the 1960s HP Sauce ran a fleet of Minor LCVs. He recalls that during the mid-1960s HP Sauce purchased two chassis/cabs and had coach built HP Sauce bottles as van backs!

There seems to be very little information on the ice cream conversions, although I have a few photographs. The companies that carried out the conversions are often unknown or no longer exist.

Mr Seniuk informed me that in 1972 he went to his local BMC agent at Kennings in Manchester to purchase a new Minor LCV and was advised that they were no longer available. He was then offered the BMC dealer's delivery van, TNA 654K, which was then six months old. He bought it and is still using it to sell ice-creams in Manchester, twenty-seven years later. The rear end of the is vehicle is aluminium, not fibre-glass, as first thought. Mr Seniuk made the rear end himself in 1960. A truly remarkable vehicle and built to last. It has currently covered 90,000 miles.

He has run four Minor LCVs ice-cream vans since 1960. The first, 3252 NC, was purchased in 1960, and had a very attractive rear body conversion, based on neither the pick-up nor the van. According to Mr Seniuk the rear end was too heavy and a further conversion was made to the same vehicle. This conversion was exactly the same as his current vehicle, TNA 654K. The livery is white with red wheel arches at the front and onto its doors. It carries Mini commercial rear lights.

Ian Gutherson, from Northumberland, who owns an ice-cream van complete with the original fridges, said that his vehicle was part of a fleet of vans run by 'Tognarelli's'. It is thought that there were two vans, both purchased new in 1957 and 1958. They were sent to Myers of Distington, in Carlisle, where the bodies were fitted. Ian's pick-up registration was TAO 211, which was sold on. It now carries YVS 329. The other pick-up was URM 620, which was thought to have been sold to a museum in Wales. Another company that converted ice-cream vans was Morrisons.

Another unusual conversion is a Bemo, which is the name given to a local bus in Jakarta, Indonesia. From photographs I have seen, it looks like a purpose-built back on a chassis-cab/pick-up. It was believed to have a wooden roof and wooden rear seats.

There were companies that converted vans to Dormobiles in the 1960s, such as Wadhams, but most of these tended only to mess about with the outsides. There are also numerous caravanette type vehicles that have been converted from other types of Minor LCVs, like gown vans, or had caravan bodies placed on to the chassis. I think most of us are aware of the Cowley Caravanette Conversion, which was designed at Cowley. I think that this is a superb vehicle; it comes complete with awning and elevating roof. (840 HCG is a good example.)

There are two vehicle conversions that were completed in the 1960s, which have until now remained a mystery. They were the Gown van and the Hi-Top van (often referred to as the

Currys van). It was thought at one time that all high-top vans were ex-Currys, but this has been proven not to be the case. We know of one such van, that was bought from a British Leyland (BL) agent in Hertfordshire, which was used as a mobile cobbler. A further two were used by laundry companies in Manchester and the Channel Islands.

Mr W.D. Pittham OBE, who was the Fleet Sales Manager who supplied Curr's with most of their cars, as well as LCVs, knows all there is to know about the Currys Hi-Top van. He informs me that during the 1950s and 1960s Marshalls were supplying most of the Currys branches with Minor vans and later with Morris 1000 vans in standard format. During this period Currys became the first agents in the UK for the IGNIS fridge/freezer. To accommodate this stand-up unit Marshall Motor Bodies, a subsidiary of the Marshall Group, designed and carried out the conversion. It consisted of a fibreglass high-top, which fitted into the van roof lip, secured onto the van internal sides, together with fibreglass extensions to the rear doors. The overall internal height was governed by the height of the Ignis fridge-freezer. This conversion remained in service for many years with Currys and continued until production of the Minor LCV ceased. I am informed somewhere between 1,100 and 1,500 were built! These vehicles also had a translucent roof panel in the fibreglass top and an extra support on the cab rear sides to add extra strength.

What is thought to be one of only two surviving gown vans has recently been purchased by Eric Payne, in Worcestershire, in virtually original condition. The gown van rear is made of a coachbuilt rear body of aluminium, with extended height and width over the standard LCV. It would have been converted outside of the factory by a coachbuilder. The rear-body section also extended over the cab and had square wheel arches. The vehicles would have left the factory as 'chassis-cabs' as the chassis number of this vehicle, NGH 97D, proves this: MAQ5 207193. The vehicle has a large, centrally-fitted rear door, which opens outwards, to enter a rear compartment of high proportions. The side and roof is lined with wood panelling, similar to that of a bus body, with two chrome garment rails that run the entire length of the body. These themselves seem to be handrails from a bus, as do the two rear air vents, mounted either side of the rear door. There are two lights, that are mounted on the inside roof, near the garment rails. These are of rounded, frosted glass on wood plinths, again very like a 1950s or 1960s bus interior.

There was also a large storage area over the cab, which was accessed from the rear, used for the transportation of fashion accessories to go with the garments or gowns such as shoes, handbags. The wheel-arches on the inside are also wood lined and edged with aluminium, with the roof being strengthened by wooden beams.

Joe Mack, the company historian at Haws-Garner, based in Andover, Hampshire, confirmed that the company carried out this conversion. The employees particularly remembered the dome over the cab. It would have been unlikely that the company would have known who the customer was. It was normal to go into the dealers and order a vehicle, then the 'chassis cab' would arrive at the coachbuilder and then return to the dealers after the conversion. Joe informs me that little has changed with these practices over the last fifty years or so. Less than fifty of these vehicles would have been converted. Many thanks to Eric Payne, in Storridge, for access to this rare and unusual vehicle, as it is currently under restoration.

Neil Ripley has informed me that bakers in the West Hull area ran a fleet of Minor vans in the 1960s into the 1970s, that had a single piece rear door, that opened upwards. It is thought that the vehicles were equipped with racking to hold the bakery trays. Neil's van, UDB 950K, was an ex-baker's van from H.S Hobson Ltd, based in the Reddish area of Manchester, and he had the racking fitted.

There is a whole chapter devoted to the unique pick-up conversions on the Mediterranean islands of Malta and Cyprus.

We now know of a number of regional newspaper companies that used the Minor LCV, but the *Daily Telegraph* is the first national newspaper that has been confirmed as a fleet user of the Minor LCV. Geoff Hoskins from Kent used to own a superb restored van painted in the colours of the *Daily Telegraph* and even managed to get them to sponsor the paint job! The 'Hull Daily Mail' used the Minor LCV to deliver its newspaper. These vehicles carried a very interesting livery of black and white, black one side and white the other

The *South Wales Argus*, a local newspaper in Gwent, ran a small fleet of nineteen Austin Minors. They were all purchased from Alsops of Newport in July 1969. The livery on these vehicles was two-tone yellow and maroon, separated by a white body stripe. A unique feature of these vehicles was that they all carried a small, roof-mounted blue light on the cab, a blue version of the 1960s' parking light accessories that was available at the time. It is not known why they needed the blue light.

David Ward from Manchester has informed me of a very rare vehicle, or fleet, run by the *Manchester Evening Mail*. They were converted and had a roller-shutter rear door. These vehicles were used in and around the Manchester area, possibly in the late 1950s to the early 1960s. There may well have been around forty or fifty of these vehicles. They had the nickname 'Pink Witches', due to the pale pink livery. I assume that the shutter-door would have been done at the factory, using perhaps part number ANK 5487, possibly from an existing vehicle in manufacture, such as the Morris J2/152.

Godfrey Crew from Newport, Gwent, has spent most of his working life running his dairy business and has used Minor LCVs in both van and pick-up types. At this time there was a company by the name of R.W. Osbourne, based in Saffron Walden, in Essex, who converted various vehicles, including the Minor LCV, into milk floats. They used to send photographs of their coachbuilt conversions to prospective buyers. I was told the cost was high but Osborne's were the 'Rolls Royce' of conversions. JDW 719F was Crew Dairy's specially ordered chassis and cab, which was to become its first converted milk float. The vehicle was left-hand drive, to give constant pavement access, as often the cab had a number of crates on the passenger seat. Cardiff Commercial Body-Shops in South Glamorgan, converted it in 1968, when the vehicle was new. He currently has a milk float conversion in his fleet, but at the present is under restoration. This conversion was carried out by Jones & Bradley of Newport, Gwent, and is based on an Austin pick-up, and carries the registration number NTH 443J.

Geoff Hoskins, in Kent, says that his father had milk float, converted by Crawford's, of Ashford in Kent. 3016 KO was the registration number of Geoff's father's van. All the vehicles are very similar and have tried to copy the coach built body of the Osbourne conversion.

Mike McGinty in Queensland, Australia, owns a vehicle once owned by Maranoa Bakeries, which later became Western Bakeries, some time in 1967. The van is now registered as BMC 60. It was first registered in 1960, with a chassis number of YJMAV3R 115014/2925. (Australian Built). The van was purchased by Max Wyatt, of Maranoa Bakeries, from the local BMC agent in Roma Amor Motors and was converted to a pie van, around 1967, and was driving around the building sites. It originally had a wood-burning oven, which has now, been converted to gas. The vehicle is also equipped with a chimney.

In the early 1990s Henrics in Nottingham made a fibreglass rear end for both the van and the pick-up, thus enabling many LCVs to continue in existence, rather than being scrapped. Nowadays you will find many LCVs on the rally field that have the rear ends in place. The Custom Fibreglass Co from Woodhall Spa, Lincolnshire, produced a rear end to convert your traveller into a van. It was rumoured that they would produce any rear end to fit your LCV! The back end fitted to the traveller bears a remarkable resemblance to the DOMI van (the Danish PO conversion), with a single piece rear-opening door.

A fleet of nineteen Austin *Argus* vans is being handed over to Ralph Jones, circulation manager of the *South Wales Argus*, by John Williams, commercial sales manager of Arthur Alsop Ltd (BMC agents) on 28 July 1969. The unique feature about this fleet was the small blue lamps on the cab roof. *Photo: J.B. Williams.*

This Morris Series V van, c.1971 is seen converted to a caravanette with elevating roof in 1989. The for sale signs are displayed on the front windscreen.
Photo: G. Crew

A row of five Currys high-top conversions, c.1960, at Marshall's Garage. Currys became the first agents in the UK for the IGNIS fridge/freezer and to accommodate this stand-up unit Marshall Motor Bodies, a subsidiary of the Marshall Group, designed and carried out the conversion. The vehicle on the far left was unregistered when the photograph was taken. *Photo: Marshall's of Cambridge.*

Ashley Watt spotted the last known roadworthy surviving high-top conversion in 1989. Sadly, it was a write-off. *Photo: Ashley Watt.*

These vehicles carried a dark-blue livery, with three white ovals with 'Currys' in red on them and 'The Friendly People' on the side panels. Between the front door and rear wheel arch was the address and telephone number of the shop to which the vehicle was allocated. This one was on its way to the store at Ashton-under-Lyne, in 1960. The extra support rail is clearly visible through the driver's window. *Photo: Marshalls of Cambridge.*

This fibreglass high-top, and the rear doors, owned by Chris De La Mare, are all that are left of a van used by a laundry company in the Channel Islands. It is now thought to be the only survivor of a possible 1,500 built. The translucent roof panel, although covered in mud, can be seen clearly in this picture from 1999. *Photo: E. Payne.*

'Check Point Charlie' and the Berlin Wall form an impressive backdrop for one of the only remaining two gown vans, NGH 97D, c.1966. The livery is all over dark-green. It is seen here in 1978. *Photo: M. Blackburn.*

Another gown van, 8099 MH, c.1960, carrying the more usual two-tone grey livery. It is thought two of fifty built survive. This example was seen on a rally-field, c.1989. I suspect the registration number now lives on, but the van has not been seen. *Photo: G. Crew.*

Bon Accord Glass, in Aberdeen, now has a specially converted Minor pick-up as part of its fleet. It carries the company colours of navy and white, c.1995. *Photo: Bon Accord Glass.*

This gypsy caravan conversion is affectionately known as the 'Hobbit', as the registration plate confirms. The van dates from 1957. It was converted by its third owners, who purchased it in 1996. There is a large stem to the wing mirror, which is needed to cope with the extra body width. The vehicle is on Waiheke, seventeen miles from Auckland in New Zealand. *Photo: B. Atkinson.*

A Swedish Post Office mail van, c.1954. A factory in Vastervik, Griparosser, in Sweden, carried out the extended body conversion. However, it is not known how many were converted. *Photo: Swedish PO Archives/ H. Pehrsson.*

D.O.M.I. (Dansk Oversoisk Motor Industrie) van, c.1962, was a special body conversion carried out for the Danish Post Office on chassis-cab imports. The only known survivor is seen here in Silkborg, Denmark, in 1996. Note the single piece rear door. *Photo: S. Marsball.*

A caravanette conversion of a Morris van, *c*.1965, at Leeds MMOC national show in 1993. At the time it was owned by Jim Lambert. The young boy looking in is the author's son James Harvey. *Photo: R. Harvey.*

A caravan company owned this caravanette conversion, with elevating roof, *c*.1970. When a delivery had been made, the delivery driver would often stay overnight in the van. It is seen here in 1989. *Photo: D. Game.*

This Austin van TOX 79H 6-cwt, c.1970, was dubbed 'The Snail', because of its caravanette conversion, carried out by L. & A. Fisher Ltd in Bagshot, Surrey. The builder's plate can be seen on the rear of the van under the window. The vehicle has been off the road since 1992. *Photo: R. Crossman.*

Western Bakeries used this Morris van, c.1960, to make deliveries to building sites at lunch and break times. It originally had a wood-burning oven fitted. In 1967 it was changed over to gas. It has a very appropriate registration number: BMC 1960. This photograph was taken in Australia in 1998. *Photo: M. McGinty.*

Tognarelli's Ice Creams of Worthington had two Morris van conversions. The vans were registered in 1957 and 1958. Until 1993 both were thought to have survived. YVS 329 is currently under restoration in the north east of England. *Photo: I. Gutherson.*

Mr D. Seniuk with his family, c.1960, in Manchester with his first Minor van. He carried out the conversion himself, but found it a little heavy, and re-bodied the van with another conversion. This has been used on another two Minor ice cream vans, including the current one, which is still working for a living thrity-five years on. *Photo: D. Seniuk.*

This is a Skinner's Ice Cream van, c.1972, from Boston, Lincolnshire. It was bought for £350 and converted in 1979. This particular conversion was completed in 1979. A similar conversion was carried out in the early 1960s. The extension, over the existing body, provided an area for serving the ice cream. At the various markets the van worked, it was reversed onto made wooden ramps to enable the vendor to serve the customers without bumping their heads! The sign was a later addition. *Photo: J. Biggs, Skinners of Boston.*

This Country Boy ice cream van, REG 527, c.1960, was part of a fleet of around twenty vehicles. It was probably converted near Bury St Edmunds. It may have only survived because of its registration number. It is seen here in 1994. *Photo: D. Estlea.*

Ex-PO Telephone engineer's van JLT 496D, c.1966, seen in 1975, was converted to an ice-cream van for a round in the Norwich area. *Photo: R. Taylor*

Godfrey Crew used his Minor van for his milk business. Here he is near his dairy in Manchester Street, Newport, in 1960. On the windscreen are popular transfers showing the driver various signs of the Highway Code. *Photo: D. Short.*

This milk float conversion by Osborne's of Saffron Walden, in Essex, was made in 1960. Osborne's conversions, with the rounded edges were the most sought after. Most other body shops tried to copy this conversion. Note the metal crates and large milk bottles.
Courtesy: G. Hoskins.

This Crew Brothers' Dairy, milk float, JDW 719F, had a body conversion from Cardiff Commercial body shops in 1968. The vehicle was purchased new as a chassis cab It was left hand drive to give constant pavement use. The rear lights are recessed to avoid damage.
Photo: G. Crew

Four
Public Utilites
(Including Local Councils and Government Departments)

Gas and Electricity

The GPO & PO was by far the largest fleet user of the Minor LCV. However, the public utilities were also large fleet users. Most regional gas and electric boards used them in large quantities, mostly in the van form, rather than the pick-up. There is little information or photos of fleets, as most of these regional boards have since been privatized and archives have been lost or destroyed.

I am thankful to vehicle enthusiasts who recorded that Minor LCVs were used by these utilities. Some also logged the registration numbers. Some photographs do exist, which clearly show the livery and boards that used them. Others give hints to what the livery was like. I have listed the known electric and gas boards that used the Minor LCV.

Eastern Electric Board (EEB) was a large fleet user of the Minor LCV, I am grateful for this information from Robin Taylor in Norwich. The livery was light-blue and dark-blue, until around 1971, when this changed to all over white. Written on the side panels was 'EASTERN ELECTRIC' in boxed, block capitals in contrasting boxes and letters. It is also known that they used at least one pick-up. It would appear that the later vans were also purchased as Austins as well as Morris. EEB used the LCVs from around 1960 and their latest van was registered on an 'L' plate in 1973. A fleet of over 200 was used.

The Southern Electricity Board (SEB) probably had the largest electric board fleet of LCVs. They had three liveries. One of the last vans used and purchased by the SEB, WRU 381K, was painted in two-tone turquoise and white, and would have had 'ELECTRICITY' written black on the side panels.

I know of at least three vehicles in the WRU registration batch, some as high as 885K, a further two others on 'K' plates and even one vehicle on an 'L' plate. The early livery of the SEB was all over dark green, with SEB, in gold, on the side panels and the rear doors. This was in use until the 1960s. The SEB took delivery of Minor vans from the start of production in 1954 and were still having Minor LCVs registered as late as 1973.

The livery used from the mid-1960s would have been light almond-green, with white front wings and a white top half of the van back above the waistline. This did not include the roof, as this was light almond-green. Written on the side panels would have been 'Southern Electricity' in black and red.

Mr Greenfield in Kent confirms that, during the 1960s and 1970s, South Eastern Electric Board, 'SeeBoard', ran a fleet of Minor LCVs; they used both vans and pick-ups. The livery of these vehicles was all over apple-green, with gold lettering and SeeBoard Logos.

SeeBoard had at least one Morris Minor pick-up specially converted. Peter Willmet from Kent has sent me a copy of an article that appeared in the *SeeBoard Staff Magazine* in 1962. It shows VAP 380, a minor pick-up in SeeBoard livery, with the 'SeeB' crest on the doors and 'ELECTRICITY' sign-written in a handwriting font. The vehicle was converted for use as a jointer's truck and used in the Lewes area. The layout of the vehicle was arranged after an investigation into use of equipment in the traditional jointer's trailers. The use of the minor pick-up and the design chosen, enables the jointer access to all his small tools and carry out a variety of small jobs with minimum inconvenience. The vehicle was also chosen to keep obstruction to traffic on busy roads to a minimum. A VHF radio was also provided in this vehicle.

South Wales & South Western Electricity Board, SWEB, used a large fleet of Minor LCVs. These vehicles were slightly different and a large number of the fleet were fitted with Minor Traveller type rear chrome bumpers. The livery carried by both companies was all over turquoise blue with SWEB written in red on the side panels. The blue was a Ford colour named Caribbean Blue from the Ford Classic/Capri at the time.

Yorkshire Electric Board (YEB) was a fleet user of the Minor van. Bill Wilkinson from Hull remembers the vans as being a livery of all-over turquoise-green, with yellow writing. Some of these vans were fitted with tow hitches and roof bars.

Wales Gas (Nwy Cymru) used the Minor LCV. They were purchased between 1967 and 1972. Alas, there seems to be no photographs of them, neither is the livery known or what duties they would have performed. The livery of these would have been all over dark green, with three white stripes, that ran behind the words written on the side panels, that is 'Gas' and 'NWY' in red, and over the top of the van. NWY appeared on one side and GAS on the other.

The States of Guernsey & Jersey Electric Companies used Minor vans during the 1960s and into the 1970s. The livery of Jersey is unknown, but Guernsey was British racing green (BRG) with black wings and wheel arches.

Back on the mainland both the Midlands Electric Board (MEB) and Manchester & North Western Electric Board (MANWEB) used Minor vans. The livery of the latter is not known, whereas MEB would have been two-tone blue and white, the dividing line running along the waistline of the van.

The only other two Gas Boards that I have been able to confirm are Southern Gas and South Eastern Gas. Liveries or numbers deployed by both companies are unknown.

Government Departments

As well as the public utilities, there were various government departments said to have used the Minor LCV. Information on these vehicles is very rare and normally extremely difficult to obtain. Certainly one of the largest known government users was the Forestry Commission. The Ministry of Defence, the Ministry of Supply and the Ministry of Works also used the LCVs. It is not known what duties were carried out by the Minor in these departments. I am also informed by both Godfrey and Richard Crew that H.M. Coastguard, H.M. Immigration and H.M. Customs all used the Minor LCV. Liveries, dates or usage are all unknown.

Forestry Commission

The Forestry Commission, now Forest Enterprise, used the Minor LCVs in both forms, pick-ups (with and without tilts) and vans. Laurence McClung, a workshop manager, now working in the Cornwall area, worked in Scotland during the 1960s and into the 1970s. He informs me that a number of Minor LCVs were sold-off at auction and went to Ireland. Laurence has confirmed the livery as Deep Bronze Green, which also included the old-style Forestry Commission logo. These vehicles also carried the royal crowns and, like Royal Mail vans, had a different crown in Scotland. These vehicles were still in active service until the late 1970s. It would appear from current records that the Forestry only used Morris 6cwt vehicles and not Austins. I have been told that the Forestry's vehicles were bought through the Ministry of Defence (MOD). JUL 537D is the only known ex-Forestry Commission Minor in preservation.

Home Office (Civil Defence)
The Civil Defence ran Minor vans on the East Coast. Two have survived. The Home Office originally purchased VLU 265G and it was registered to the Secretary of State. It was part of the Emergency Fire Service (EFS) as a fleet support vehicle to the fleet of Green Goddess's. The EFS covered early warning systems for nuclear attack, water supply, flood control as well as fire engines. The original location of 'VLU' is still a mystery. It would have been in a stores site where some of the E.M.S. was maintained. The vehicle then lay unused for a number of years and then ended up in Burton-upon-Trent, where it was kept in a heated garage, covered and up on blocks.

In 1987 'VLU' was lent to the Police Training School at Newbold Revel. In 1993 it was sold to a private dealer. A Mr Walker bought it in 1994 and the vehicle is currently under restoration. In 1994 it had less than 45,000 miles recorded on the clock.

Local Councils
Local Authorities up and down the UK have used the LCV for a wide range of duties in various council departments from building repairs to meal on wheels. It was also not unusual to have different liveries for different departments. During 1974 the water services were passed from local authority care into the hands of the water authorities, which often resulted again in a change of livery. Also, in this year borough and district councils were formed and new counties were formed, like Avon and amalgamated, like Herefordshire and Worcestershire

A number of local councils in the London area were fleet users. From 1964 the Greater London Council was formed with the present London boroughs, which would have involved change of ownership and liveries. Ealing, had a livery of white and maroon, Hammersmith was orange and black, Islington was green and Westminster was maroon. I am sure that there are many others.

Hull City Council were users of the Minor LCV and, as already mentioned, may have claim to the newest van in existence. The livery carried by Hull was yellow and blue with the crowns not unlike their telephones van.

Many local, regional and district councils all over the UK had fleets, far too many to list here, but some notable ones are as follows. York City Council had a 'pest control van' run by the environmental health department, which carried a livery of fawn, with maroon trim and a three figure fleet number.

Worcestershire County Council used them as part of their highways' division, which also included Austins. The liveries carried were either light-green, or orange for the highways' division. Cardiff City Council was a large fleet user. It ran a fleet of over forty-five Minor LCVs. Most departments were in different liveries, including green, blue and red. They transferred vehicles over to South Glamorgan County Council when it was set up in 1974.

Over the water from Cardiff to Bristol where again the council found a use for the van, they carried a livery of light-grey.

Jersey Parish Council managed to find a use for the Minor pick-up, whereas Essex County Council used a pick-up as a grounds-mans vehicle. The livery carried here was yellow with red and black lettering. Oxford City Council even used a Minor van as a lamplighter van; this vehicle still exists to this day in private hands.

Manchester City Council, Norwich City Council, Surrey County Council, Hereford City Council and Gloucestershire County Council used the Minor LCV and I am sure plenty of others did as well.

Water Authorities

Bristol Water currently have a restored a Minor van. Victor Skuse, the transport manager of BWW was responsible for the restoration of the vehicle. The vehicles is not alas a genuine BWW van but has had a full restoration and carries an original BWW registration number 43 MHY when exhibited at shows. This vehicle carries what is believed to be the first livery carried by the Minor vans. This one is dark-green with black & yellow chevrons on the rear doors. It also has BWW in gold on the side panels.

The second livery is a two-tone affair consisting of Marine Blue and Light Admiralty Grey the waistline of the van is the dividing line of the two colours. There is a story that a van displayed both liveries on either sides of the vehicle, but unfortunately no photographic evidence exists.

The fleet was more than fifty with at least one known to be a pick-up complete with canvas tilt and most carried an orange light.

Another Water Authority fleet user was Ayr Water and it would appear in May 1961 the Ayr Water Committee authorised the purchase of a Morris 5cwt Van and the cost at that time was £392 15s. Hull Water Board used the Minor van, but little else is known other than the livery was blue and pale-blue.

The states of Guernsey Water Board used the Minor van. The livery was cornflower-blue. East Suffolk & Norfolk River Authority were also users of Minor vans, these duties may have included catching poachers and catching people fishing without a licence, the livery was all over cream with black lettering. Also the South Norfolk Water Board used some splendid liveried vans in pale blue with green and white lettering.

Other confirmed Water Authority users include the Anglian Water Board, with vehicles transferred from Smallborough Regional District Council. The livery was white and blue. The North Yorkshire Water Authority used an almond-green livery, whereas the Rickmandsworth and Uxbridge Valley Water Company was all over dark green. The Isle of Wight Water Board also ran a small fleet but alas the livery is unknown.

British Waterways also had Minor Minor vans. At least two vehicles were still in service up to 1976 or 1977.

A Forestry Commission pick-up, YYY 684G, c.1968, complete with canvas tilt, near the reservoirs at Brecon in 1971. *Photo: K. Porter.*

Home Office Minor van VLU 265G, *c.*1968, was part of the Emergency Fire Service (EFS), as a support vehicle. The EFS covered early warning systems for nuclear attack. Here it is shown in 1994, when it finally fell into private hands. The vehicle is in an all-original condition, with its livery of dark green. The bonnet badge, for some unknown reason, is plated over. *Photo: G. Crew.*

It is known that the Army used the Minor Traveller for bomb disposal. However, details of military vehicles are always difficult to obtain, so perhaps they did use vans. If they did, is this what they would have looked like? This particular one was restored by Godfrey Crew in 1999. *Photo: G. Crew.*

An Eastern Electricity Board Austin van, PCL 485J, c.1971, in the Norwich area in 1977, just after it was taken out of service. *Photo: R. Taylor.*

Eastern Plants gave this ex-Anglian Water Board Austin van, MDX 840G, c.1968, a new lease of life. It was used as a service vehicle *Photo: D. Game.*

A large selection of commercials used by the Southern Electricity Board (SEB), c.1965. They include Morris Js, Morris J4s and at least four Morris Minor vans. Note the deep shine on the Minor van closest to the camera. The roof aerial indicated that two-way radios were installed.

SEB Minor vans form this line-up from 1960. Three of them are Series Is. All four have 'bats ears' indicators fitted on the van backs. *Photos: SEB Museum.*

73

Both liveries on these Bristol Water Works (BWW) Morris 6-cwt vans, c.1967, were carried by the water board. Each vehicle had a two-way radio fitted.

The BWW Minor van was the first company vehicle to be fitted with a night-light. The WATER sign was illuminated by a fluorescent tube, which was operated when the sidelights were switched on.

Five
Police Minor Vans
(Including Military and Emergency LCVs)

The Police force in the UK used the Minor in all its forms, including the Minor LCV van. The constabularies mentioned are all confirmed fleet users of Minor vans. It is also thought that the Derbyshire force used them, but there is no confirmation of this. According to the Metropolitan Police Museum, the main uses for the Minor LCV would have been as dog-vans, scenes of crime and incident vans. It is unlikely that they would have been used for observations or surveillance, as they were too small, but I am sure this was not the case.

Graham Kinsman informs me that the Avon and Somerset Police also had at least two unmarked vehicles. The first was a white unmarked van in the Bath area. It was still in active service until 1978 or 1979. This vehicle was used as a 'speed trap'. It was easily recognized as the van was usually parked on the roadside and needed to have the rear door open. About 100 yards away was constable on a two-way radio. The vehicle was well known in the area! The second van was used in the Bristol area. It was used in the city to set up various speed traps. This vehicle was thought to be brown, probably rose-taupe. Obviously, neither of the vehicles carried blue beacons. This vehicle was still in active service in the late 1970s.

Avon and Somerset Police, special branch, inherited a light brown (probably rose-taupe) van which was used for 'door-to-door' enquiries. They were also used as dog vans. The size of their fleet is unknown.

Colin Chipperfield said that the police in Bermuda used the Minor van, which was royal blue with the cab-roof and main body roof in white. The size of the fleet is unknown. The vehicle also carried a single blue beacon and a chrome bell mounted on the front off side bumper. They were mainly used as dog-vans. They also carried the Royal Bermudan Crest just behind the front doors; the only registration available is JX 181.

Steve Woodward confirms that the Berwick, Roxborough and Selkirk Constabulary had a fleet of minor vans. They carried an all white livery, with 'POLICE' in blue on the van side panel. There was also a small 'POLICE' sign in blue and white, on the cab roof. The only registration number available is '969 PDS'.

Cardiff City Police, now known as South Wales Police, used Minors as dog-vans. They carried a livery of all over black, with 'CARDIFF CITY DOG SECTION', in white, on the side panel and the city crest beneath it. The vehicle also carried a blue beacon and a roof mounted air vent at the rear. The only registration number available is FUH 167C.

South Wales police in the Cardiff area used an unmarked grey van. It was used by the scenes of crime and the forensic department. Details on this one are a little scarce.

South Wales Police (SWP) archives have managed to trace and establish four Minor vans. These were 316-318 MTG. All three were bought in November 1961 and sold between 1965 and 1966 for £70-£100, with around 74 to 80,000 miles on the clock.

Mostly the vans were used as dog-vans. There appears to have been two liveries, possibly due to the various constabularies before the merger to form SWP They were all over white with 'Police' in black (mainly dog-vans), or all over navy blue, with 'Police' in white. SWP also had a 'specialist searches' van, using a Labrador for 'drug searches'. The vehicle was painted maroon and would have been unmarked.

According to the Police Museum at Bridgend, Minor dog-vans had a hand lamp with a powerful beam for use in connection with the examination of buildings. In 1967 to 1968 they would have cost £4. It is worth noting that on 27 November 1967 an order was placed for seventy-seven Minor Panda Cars.

Bob Edwards in South Wales has what is thought to be an ex-Police dog-van. He has owned the vehicle since 1970. It was just two-years-old when the Police disposed of it. The van had covered some 60,000 miles in active service. It was used in the Barry Docks area, during the late 1960s as a patrol-come-dog van. It is thought that the Barry Police may have inherited the van, as it was unmarked and its livery was rose-taupe. Since Bob has owned the vehicle it has given him very little trouble. It has never been welded. He purchased the van from a garage in Nantgarw for £265. The van still has the holes where the blue beacon light unit would have been fitted in the middle of the roof of the van-back. The van has now covered over 120,000 miles and has been to Lands End and John O'Groats on several occasions. Not long after Bob purchased the van, he was waved down by the Police, only to discover that the vehicle was an ex-police van. The policeman used to drive it in the Barry area. He also informed Bob that there was another identical van! Alas, the registration number of the other van is unknown. Bob's van is PTG 60F and the chassis number is MAV5 231186.

Merthyr Borough Police used an unspecified number of LCVs from 1968 onwards. The livery was blue, with 'POLICE' in white, on the side panels. A photograph at the SWP Museum in Bridgend shows an LCV with registration number CHB 750G.

Cumbria Police also had minor vans. These vehicles were white, with 'POLICE' on a blue sign on the van side panel. The vehicles also carried a roof box with a blue beacon. However, no registration numbers are available.

East Suffolk constabulary used the Minor LCV as part of their traffic or motorway patrol vehicles. On the rear nearside door is a 'POLICE STOP' sign, around twelve inches square, which would have been illuminated. The vans were white, with 'POLICE' in blue, on the side panels.

Gateshead Borough Police used the Minor van as incident-vans. They were dark-blue, with 'Gateshead Borough Police Incident Van' in white, on the side panels and a single blue beacon. The only registration number that we have is NFV 744.

Brian Homans, of the Police Vehicle Enthusiasts Club (PEVC), has confirmed that Hertfordshire police used Minor vans as far back as 1955. The vans were navy-blue and unmarked. TNK 530 was used by the photographic department. Hertfordshire also had an unspecified number of Travellers in the 1960s, but it is unknown if they were dog-vans.

In the late 1960s and early 1970s, Humberside Police used unmarked, aqua-green vans in the Hull area. They were used for speed traps, by the CID or the scenes of crime department.

Glasgow Police, now part of the Strathclyde Constabulary, used the Minor LCV as Panda cars. The CID used brown travellers. The livery was pale-blue with white doors and cab-roof, not unlike some of the Pandas and Travellers used by other constabularies at this time. They had some unique features: bumper mounted twin spotlights, a roof aerial mounted on the van-back and black wing-mirrors. The registration numbers are in the SYS xxxG series. Strathclyde Police was formed in 1975, after the amalgamation of the forces of City of Glasgow, Dumbartonshire, Renrewshire, Lanarkshire, Ayrshire and Argyllshire, all of which used the minor saloon, traveller or van and in some used all three.

Hampshire Police used the Minor LCV van. Steve Woodward confirms that GHO 407D was originally purchased by Portsmouth City Police in 1966 and was retired in February 1971, having covered 98,120 miles. It was sold at Frome car auctions for £145. GHO 406D was bought at the same time and retired a month earlier in 1971, but this vehicle had covered 145,165 miles. It was also sold at Frome car auctions for £140. Both vehicles were mid-green and were used as dog-vans. They had two roof mounted police signs on the cab and on the back box section and a blue beacon. WOR 670, 870 MAA, 830 EAA, 831 EAA, 114 BOT and XDL 858 were all registered with Hampshire. Very few details on these vehicles exist and it is not known what duties they were involved in.

Chris Bowkett confrims that Kent Police used unmarked, brown, possibly rose-taupe, Minor vans. They were assigned to the traffic department they were parked on the bridges over the M2. They had more aerials than a Russian Sputnik.

In 1964 Leicestershire Police used two Morris 1,000 6cwt vans on the M1, as emergency vehicles. The vans were fitted with a two-way radio and a complete set of equipment for scene of crime or accident situations. The vehicles were run by the CID photographic department. and were manned by two officers. They were white, with blue side panels and 'POLICE' in white. They also had a crest behind the front doors, in front of the wheel arch. They had a roof mounted: blue beacon, a four-foot aerial and 'POLICE STOP' sign over the rear doors. Mounted on the cab roof were an illuminated 'POLICE' sign, siren and a P.A. system. In the back were signs, warning cones, axes, cutters, crowbars, ropes and lamps.

Ray Seal of the Metropolitan Police Museum confirms that the Metropolitan Police also used the Minor LCV in large numbers, as dog-vans. The registration numbers would have been part of a batch in the JGY xxxK series. JGY 682K, JGY 688K, & JGY 689K are all confirmed. They all have a circular air vent, a blue beacon and aerial on the rear of the roof. They were dark blue, with 'Metropolitan-Police'on the side panels. According to Bob Appleton, a surveillance photographer with the 'Met' from 1974, the photographer could choose any type of 'covert observation van', providing that it was built in the UK. Bob inherited his light brown Minor 1,000, MGK 360L. The vehicle was unmarked and would have had darkened windows, a screen behind the driver and perhaps some spy holes. It was still in use in 1975. The 'L' on the plate denoted another usage.

Fife & Perth Constabulary used the Minor van from around 1959 until the early 1960s, according to Ken Taylor from St Andrews. He remembers using one to help him move in 1959, just after he got married! These vehicles were used as rural 'Pandas'. They were black with 'POLICE', in white, on the side panels. The registration numbers were in the sequence RFG 2xx. Minor vans were also used in the Perth and Kirkcaldy area. Their livery and duties are unknown, it is thought that the registration numbers were AXA-A, an unused registration batch from the London area.

Colin Chipperfield, a member of PVEC, is the man whose photos of Minors appeared in 'The Sun' a couple of years ago. He has over 15,000 photos of police vehicles. and has confirmed the Thames Valley Police used Minor vans, probably as dog-vans.

West Yorkshire constabulary used the Minor van mainly as dog-vans. They were navy blue, with 'POLICE', in white, on the side panel and DOG PATROL, in white, below the waistline, under the side panel. 2996 YG was fitted with a roof-mounted air-vent, and an aerial and blue acorn on the van-back roof. It also had a spotlight on the offside on the front bumper and black wing mirrors.

Steve Woodward confirms that Wiltshire Police also used the Minor LCV. Their livery was green and yellow. The vans also carried a blue-lamp, with a 'POLICE' sign mounted over the back doors. 'POLICE' in black was written on the yellow side panels They used LCVs instead of Morris Minor Panda cars on a 'Unit Beat Police Scheme', known as the Rural Beat Scheme. Fleet numbers are unknown. The only registration number available is VHR 224K.

Military Police (Royal Navy)
The Royal Navy used the Minor LCV during the 1970s, according to Richard Crew, in Gwent. The Royal Navy military police used them in and around the Portsmouth area, until 1975. Fleet numbers and duties are unknown. The livery was navy-blue with 'Royal Navy' in white, on the side panels.

St Andrews Ambulance
During the mid-1960s the Scottish Ambulance Service (Scottish version of St Johns) introduced a pilot scheme involving mobile attendants with Minor vans. The only known registration number was LUS 933E. The livery was all over mid-dark blue, with sirens fitted as standard. They were equipped with two-way radio-telephones and blue beacons. Initially, the vans carried spare oxygen and blankets. They were gradually upgraded until they became fully equipped emergency vans in 1968. Some even carried extrication equipment, a duty now performed by the Fire Brigade. In 1969, all remaining Morris 1,000 vans were re-sprayed off-white and became Mobile Incident Vehicles. When attending a major incident, a flag bearing the Saltire Cross displayed on a flag base, welded on the van's roof.

Fire Service
At least two Fire Brigades, Buckinghamshire and Gwynedd, used the Minor LCV They would have been used as a run-about between fire stations or to visit public buildings to check on fire safety. The livery was usually red, with the county name in gold. Some would have a blue beacon fitted.

Ex-South Wales Police dog-van, c.1967, seen here in 1997, now in the hands of Robert Edwards from South Wales. *Photo R. Harvey.*

These Metropolitan Police Morris vans, c.1972, were used as dog-vans. They were numbered in the 'JGYxxxK' series. The vans shown are brand new. Note that each has an air vent on the roof. *Photo: R.Seal, Met Police Museum.*

The traffic division of the East Suffolk Police used this c.1969 van. There is a 'POLICE STOP' sign on the rear door. *Photo: PM Photography.*

This is an unmarked surveillance van, c.1973, used by the Metropolitan Police. Although it is only two years old, it has a number of bumps and dents. No light can be seen through the driver's window from the rear doors, indicating that a blackout screen was fitted behind the driver's seat. *Photo: R.J. Appleton.*

This Wiltshire Constabulary Minor van, c.1973, is in the striking livery of green and yellow used as an alternative on duties that would have normally been carried by a Panda car. *Photo: S. Woodward Collection.*

A Minor quarter-ton dog-van, c.1960, from the West Yorkshire Constabulary. *Photo: C. Chipperfield Collection.*

A Series II Morris quarter-ton van from the Hertfordshire Constabulary photographic department, seen at an incident on the A5 at Markyate, Hertfordshire, on 26 May 1955. *Photo: B. Homans (PVEC).*

Royal Bermuda Police Force van JX18, c.1965, used as a dog van. Note the chrome bell on the chrome bumper. Unfortunately, neither the dog nor the dog handler's names are known. *Photo: C. Chipperfield Collection.*

A Minor dog-van used by the City of Cardiff Police, c.1965. *Photo: S Woodward Collection.*

The striking livery of this new 1969 City of Glasgow Police van has brought the station staff out to have a look. *Photo: C. Chipperfield Collection.*

This Buckinghamshire Fire Brigade Minor van, c.1966, was seen in Slough, Bucks, on 26 August 1972, before it became Berkshire.

A Bangor Fire Brigade Austin Minor van, c.1968, on 20 August 1973. *Photos: K. Porter.*

Six
Malta and Cyprus

While holidaymakers come to Malta and Cyprus in search of a Mediterranean suntan, enthusiasts of classic British cars flock to the islands to search for Morris Minors from the 1960s. This of course includes the Minor LCV. Although they are now a rare sight, Series IIs can be seen working every day on both islands. Most are in good condition, which is largely due to the climate. There seems to be little information on any fleet users of the Minor LCVs on either island, but many tradesmen, including ice-cream vendors and estate agents found a use for the Minor LCV.

A Cyprus petrol company used a fleet of at least four vans, according to John Baker from Preston. There are two blue vans, in a scrapyard near Larnaca., with ΠΕΤΡΟΛΙΝΑ, in white, on the side panel. Andy Ballisat recorded the same vehicles. They were owned by Petrolina Ltd. The livery is Persian blue, with black wings front and rear. On the nearside panel is 'ΠΕΤΡΟΛΙΝΑ' and on the off side is 'PETROLINA' in white. At least one of these vans is an ex-GPO van, complete with roof ladder bracket and pigs-ears indicators.

Maltese Pick-ups

Maltese type pick-ups have a unique rear-end. I am indebted to John Vella from Malta for a number of photographs and information on the LCVs found on Malta. Pick-ups were imported to the island, as second hand from the UK. Often the cab-backs were removed before export. This was done for two reasons. Firstly, to import commercial vehicles into Malta there was a 65% customs duty to be paid, but for chassis/cabs it was a mere 15%! Secondly, when shipping them to Malta they had to use 40ft containers and obviously with chassis/cabs you could import more.

The pick-up rears were manufactured on the island. LCVs that came to Malta this way had heaters, whereas commercials that were imported as new were supplied without heaters, due to the Maltese climate. This also explains as to why the pick-up seems to out number the van by about fifty to one There are two different types of pick-up rears. One has a small body-skirt, of two to three inches, at the bottom all the way around. There are three different types of rear light: original Minor LCV lights, Mk I Mini rear lights and a third type from an unknown vehicle. The rear-end also has three different types of tailgate. The first type is flush, the second is made from two panels. The third is a drop-sided pick-up, based on the standard 'Maltese' rear-end, but fitted 'drop-sides and tailgate. All these pick-ups have a single piece, rear bumper, possibly a piece of angle iron.

Cypriot Pick-ups

The Minor LCV features in both northern and southern Cyprus. Many are still used daily by their owners. The Cypriot pick-up is totally different to the Maltese one. It has a standard pick-up rear and a three-sided metal cage on the pick-up rear. The cage fits on to the pick-up sides. The two sides of the cage are made from four vertical stays crossed by three horizontal stays, which are linked together by a horizontal stay across the van cab-back. These cages are attached utilizing the hoop holes in the rear of the pick-up. None of the pick-ups have tilts. They must have been exported with open backs. Both vans and pick-ups from Series II to the later Series V vehicles are still found working on Cyprus. The island has become a spotters' paradise, for LCVs, second only to Malta.

John Ross, a teacher at Logos School, Limassol, Cyprus, is pictured here with his wife Jennifer and daughter Rachel, in 1994. The van U197 is a Series II, c.1955. *Photo: J. Ross.*

The body to this Maltese style pick-up, EAB 221, was produced locally on the island. It had squarer wheel arches and an extension, over a standard pick-up. It is shown here in St Paul's, Malta c.1995. *Photo: C Letton.*

This Cypriot style pick-up has the standard pick-up rear, but a cage, produced on the island, has been added to the rear. This example was seen in 1999 in Paralimni, southern Cyprus. *Photo: A. Ballisat.*

The rear end of this Maltese pick-up, c.1995, has locally produced plastic tilt covers. The lights fitted are off the early mini or commercial range. This tailgate is panelled, but a number are smooth. There is an MO Traveller parked in front of the pick-up. *Photo: J. Vella.*

The rot got so bad in this Cypriot pick-up that the sides were just cut down. The vehicle was still in use in 1992. *Photo: B. Carter.*

Local tyre fitters on Cyprus find a use for a Minor pick-up. The 'cyprus cage' comes in handy for piling up the tyres in 1999. When full, it is just driven away to dispose of the tyres! It is also equipped with Bibendum on the roof. *Photo: A. Ballisat.*

This Minor pick-up, complete with the Maltese rear end, was being used as transport for a market stallholder selling fresh local produce in 1995. *Photo: J. Vella.*

Wearing a large blue sombrero, this Cypriot pick-up was seen, in 1999, cruising or parked near the beach, near Larnaca, advertising the Azectas Mexican restaurant. In the evening it is parked near the restaurant. *Photo: J. Baker.*

This Maltese pick-up was found still hard at work at the harbour in 1995. The owner is waiting for the boat to arrive. Then the fish can be sent to market. The rear of the pick-up is protected from the hot sun by a wooden board. *Photo: A. Probyn.*

This Morris quarter-ton van, JAJ 035, *c.*1960, was seen on Malta in 1994. A genuine van is not that common, as the Maltese pick-ups outnumber them. This van has had a one-piece side window conversion and a home-made rear bumper. Typical Maltese houses, with verandas, made from the local sandstone, can be seen in the background.
Photo: Minor LCV/Dave Thomas Collection.

This working Morris 6-cwt van, *c*.1963, is resting in the midday Cypriot heat, by the look of the shadows, in 1994. It is advertising cooking oil. *Photo: LCV Register/D Thomas Collection.*

Morris 6-cwt Maltese pick-up, with the moulded skirt at the bottom of this rear-end, seen in 1998. It has a plastic tilt and an unusual bumper from a different vehicle. *Photo: A Ballisat.*

Seven
Roadside Assistance

Automobile Association (AA)

The AA used the Minor LCV for roadside assistance from the late 1960s to the early 1970s. They had a fleet of sixty to eighty vehicles. Most are thought to be Austins. However, UGN 917F was the exception to the rule.

UGN 917F seems to be the oldest Minor LCV that the AA deployed. At first it was thought that it was allocated to the Scottish Highlands, as a demonstrator. That is why it was a Morris not an Austin. All records held by Michael Passmore, at the AA archives, have the vehicles listed as Austins. However, some ex-AA patrolmen have produced photos of their vehicles, some of which were Morris vans. They all had a tow-bar, twin spotlights, one spot and one fog. The rear number plate was roof mounted over the rear doors to allow for the tilting of the tow-bar and unrestricted access when towing another vehicle. There was also an orange light on the roof vehicles and black door surrounds. The AA signs, side by side block letters, were used on the side panels and rear doors. There was also a cab mounted 'AA Service' board. There was also a two-way radio. The livery was AA buttercup yellow and was applied specially at the works.

The AA Archives have now purchased an ex-AA Morris van, YKK 643H. The intention is to get a team of volunteers, led by Trevor Groves, to restore the vehicle. It was discovered that three ex-AA vehicles were on the LCV register and the AA had to have one. There is a possibility that YKK 760H will be repainted in AA colours by its current owner. The AA also had another van that was painted red, instead of the normal buttercup yellow. It was fitted with a Pye two-way radio. It was used as a support, stores and service vehicle for the fleet in the Dover area.

Ex-chief inspector George Brown, now retired and residing in Western Australia, who covered Scotland-North, informs me that he used UGN 917F, which he thought was the only Morris Minor van that the AA had in the Highlands area. He was both lucky and unlucky with this vehicle.

On a stormy night returning from Glasgow to Inverness, in very heavy rain, he ran into the back of an army lorry that had broken down and was parked without any lights. He had to pull out to avoid a pedestrian and, on returning to the correct side of the road, the outline of another lorry could be seen, but it was far too late to stop. George had only had the van for a couple of weeks. It was the van's first accident. However, it was very lucky for George, as the rear chassis of the lorry dug into the high bonnet of the Minor, which saved him from going under the truck. He had to kick the door open to get out of the van. An army officer made the following comment 'Oh, an AA man, I have always been meaning to join!' You can imagine George's reply.

According to Mick Roberts from Eastbourne, a retired AA patrolman who drove the Minor LCV breakdown vans in the 1960s and 1970s, most of the AA vehicles were purchased from Godfrey Davies, on the Fulham Palace Road in Richmond, Surrey, in the 1960s and 1970s. Also based on Fulham Palace Road, at number 161, was a large AA garage where a lot of the AA extras were added after delivery. Dark blue Olympic radios were fitted. Another feature was a large wooden toolbox, that was placed behind the seat and a low wire-mesh screen. Mick's original toolkit, from his Minor, will be used in YYK, when the restoration is completed.

Many retired AA staff report that the AA's Austin Minor vans were the end of a large order not fully taken by the Post Office. It is not clear how true this was, bearing in mind that most of the AA's fleet were 'H' plates and the Post Office were still taking deliveries up to 'K' and beyond. Some were even registered on 'M' plates.

Royal Automobile Club (RAC)

It is not known how many Minor LCVs were used by the RAC. They would have been used for roadside service, just like the AA. RAC also has a restored Austin van, CMH 62H, but alas not a genuine one. RAC vans were RAC blue, with white side panels and a white roof. They carried a roof-mounted RAC Road Service board on the van back. Archive records also show a new RAC badge, no roof board, but an orange light on top of an illuminated box, with RAC on it, similar to the POLICE sign that appeared on Panda cars. All vehicles had hubcaps and were all thought to be Austins, not Morris'.

640 PMM, is RAC blue with 'ROYAL AUTOMOBILE CLUB' above the windows, on the back doors and along the van waistline, just under the side panels. On the side panels and the raised panels, on the rears doors, are cast RAC diamond shape plates, secured by four bolts. Some of these vehicles would have been deployed on sign duties, that is erection and removal of RAC temporary signs for large events.

The Minor LCV was used by Australian roadside assistance associations in the 1960s. each of the six states and two territories in Australia has its own roadside assistance association. They are equivalent to the RAC and the AA.
South Australia: Royal Automobile Association of S.A. Inc. (RAA)
New South Wales: National Roads and Motoring Association (NRMA)
Victoria: Royal Automobile Club of Victoria (RACV)
Queensland: Royal Automobile Club of Queensland (RACQ)
Western Australia: Royal Automobile Club of WA (RACWA)
Tasmania: Royal Automobile Club of Tasmania (RACT)
Northern Territory: Automobile Association of Northern Territory (AANT)
Australian Capital Territory: Come under New South Wales (NRMA)
Both the NRMA and the RAA used the Minor LCV.

Royal Automobile Association (RAA Australia)

Like GPO telephone vans in Britain, the RAA vans were allocated to patrolmen and were kept until mileage reached around 40,000 or the vehicle was five years old. Vans were also rotated amongst patrolmen, if the mileage on one was low and vehicles were continually renewed. Between sixty-five to seventy would have been in the fleet at any one time.

The keepers of the vans would have had them serviced; changed the oil every 1,000 miles and greased them every 500 miles. All the RAA vans were fitted with a battery charger. They were always charged-up when not in use, by plugging into the power supply in their homes.

These vehicles also carried a number of features unique to their fleet. They had large one-piece windows in place of the side panels. There were also two flat roof vents on the van back rear. The vehicles also had some minor electrical wiring modifications, to accommodate some extra lights, including the RAA illuminated globe, which was located in front of the air vents on the van back roof; the two-way radios and a second battery. The vehicles also had external sun-visors fitted, as well as petrol gauge dipsticks, as it was rumoured that the gauges were not accurate enough, and double-sided wheel nuts, like GPO vans in Britain. The dipstick was calibrated in gallons. It was accessed via a trapdoor cut in the wooden floor. The patrolmen had to put in a monthly return for the amount of fuel used.

The livery was RAA yellow, which was stamped on the Australian compliance plate on the bulkhead. The vehicles were then hand sign written, in dark blue, outlined in black, under the side windows. Prior to reselling, the vehicles would have had all the RAA special equipment removed and repainted, before being offered for public tender.

John Jones, patrolman number sixty-four with the RAA, who started back in 1964, recalled an amusing story. "I just remember the little darling with great affection. It went well and did whatever was asked of it, twenty-four hours around the clock, in all weathers. No heaters or air con in those days! Probably the fondest memory I have, is returning to Colley Street, a small street at the mar of the RAA Headquarters of the time, at 04.00hrs in the morning, flat out in second gear and then buttoning off the accelerator to an exhaust crackle and backfire, that only a finely tuned Morris Minor can do (and all of the patrols seemed to do this!). How the residents put up with this noise echoing off the brick walls of a narrow lane every night I will never know."

When these vehicles were sold off they often went on to lead a second commercial life, just like ex-GPO vans in Britain.

National Roads and Motoring Association (NRMA Australia)

The National Roads and Motoring Association (NRMA) were fleet users of the Minor LCV, based in and around Sydney, New South Wales. They offered a roadside assistance service like the RAA. The NRMA used standard vans, without any rear windows, which seemed to be unusual for Australian vans. They started using the vans from the early 1960s and they were working into the 1970s. The livery was navy-blue with a white roof and an 'NRMA. Road Service' sign written on the side panels. The vehicles also carried a large illuminated fourteen inch badge on the cab roof and a draught excluder was fitted to the driver's window. Between the rear wheel arch and front doors 'NRMA Sydney' and the address was painted. They had chrome hubcaps and silver painted wheels. The side panels also included display boards that encouraged 'JOINING NOW' and motoring security. The vans were kitted out with the tools of the trade in the rear. They also carried pig's-ears indicators, but placed uniquely on the back roof towards the rear corner. Two-way radios were also fitted to the vehicles and the aerial was mounted centrally in the roof of the van back. The size of the NRMA is not known.

An RAC publicity photograph, from 1997, showing their newly restored Austin Minor van CMH 62H. *Photo: RAC Archives.*

An AA patrolman, with his AA Austin Minor YYK 526H, is attending to a Morris 1100/1300 in 1972. The van has its number plate fitted over the rear doors, because of the tow-bar. It also benefits from town and country tyres on the rear. A Cortina Mk III can be seen on the left. *Photo: AA Archives.*

AA patrolman Steve Rymer, in Hounslow with his new Austin Minor 8–cwt van in 1969. Twin-spotlights have been fitted. *Photo: AA Archives*.

At one time this was thought to be the only Morris Minor van that the AA purchased. AA Morris van, *c*.1968, UGN 917F was not the only Morris van, although most AA vans were Austins. The patrolman is seen here giving directions. *Photo: AA Archives*

This RAC line-up in 1973, features most types of commercials used by them at this time. The Minor van shown is an Austin 8 cwt. The other small vans noted are a Vauxhall HA, a Ford Escort MK1, a Mini van and the Minor's successor: the Marina. *Photo: RAC Library*

An AA patrolman stands next to his Morris van 'EGF 622F', while he is repairing a windscreen wiper, in the early 1970s. *Photo AA Archives.*

An RAA 1960, seen here in Adelaide with its patrolman, 'Skeeter' Hassett in 1964. The van was fitted with the large side windows, which were installed when the vehicle was new. Note the external sun visor, which was a standard RAA specification. *Photo: RAA Archives.*

An RAA petrol tank, showing the dipstick that was used. The notches indicate gallons. The dipstick was secured by a screw thread into the tank and was accessed via a trap door cut in the wooden floor. *Photo: S. Holmes.*

This Ex-RAA, owned by Lesley Bourman, is awaiting restoration in Adelaide, South Australia. The vehicle found a second commercial life with 'Cooper Auto Electrics', after being sold off. Note the two air vents on the roof. *Photo: L. Bourman.*

The palm trees in the background give an exotic feel to the Sydney landscape, for this NRMA Minor quarter-*ton van*, c.1960, attending a 'call-out' for a Holden FB. *Photo: NRMA Corporate Archives.*

Eight
Minor LCV Register and Current Vehicles (Including Film and TV)

The Minor LCV (Light Commercial Vehicle) register was set up in 1992 by Dave Thomas, to cater for the Morris and Austin vans and pick-ups. There are still many thousands of Morris Minor cars around the country. However, the commercials were the real workers and consequently had hard lives. Most firms managed to find a use for them. Up to a couple of years ago they were still the poor relation and had become quite rare. This register has been set up specifically for them.

At the last count, there was approximately 3,500 known survivors worldwide, out of a production run of over 327,000. Many are now being completely rebuilt as either working 'adverts' or for everyday use. Unlike the cars, they virtually always have to be painted back to original colours, when rebuilt. The commercials can be rebuilt to their new owners' specifications, as there are no restrictions on finishes and guises. There is nothing more pleasing to see than a working van or pick-up that is painted up in the owner's business colours. There is also an Austin version of the commercial, which is less well known, as it was only produced from 1968 to the close of production in 1971.

The Great Dorset Steam Fair is the annual meeting place of the Minor LCV Register, They have been exhibiting since 1993. We are still amazed at the interest in these loveable old 'bone rattling workers'. There is a constant and steady stream of visitors who come to the stand, usually with the opening words "Cor, I learnt to drive in one of these!". Details of the Minor LCV Register can be obtained from the address below. Please remember to enclose a SAE.

Dave Thomas
Minor 1000 L.C.V. Register,
61, Zeals Rise, Zeals.
Warminster.
BA12 6PL.

The Minor LCV Register also attends the 'Austin Morris Day' that normally takes place at the Brooklands Museum, in Surrey, during February. It is organised by museum curator, and LCV member, Julian Temple. This involves a procession onto the old Brooklands racetrack. There are also meets in the North West, held at RAF Woodvale, in Southport, which is organized by another LCV member Jim Murphy, and in the Midlands at Avoncroft Museum of Buildings, which is organized by LCV members Brian Lee & Eric Payne.

There is great demand from small firms and self-employed for Minor LCVs. They are often painted to advertise the company, its products or services.

Many owners are LCV members, but a large percentage are not. When asked why they have a Minor LCV, virtually everyone has memories of the Morris Minor and its reliability. Also, they are unusual today, so they always get noticed.

Mr N. Aresti, an LCV Register member from Cyprus, has recently had a van restored in the livery of his estate agency. The vehicle is parked everyday on the sea front in Paphos. It is referred to as his twenty-four-hour salesman!

Up and down the country firms still use Minor vans and pick-ups. 'Bon Accord', a glass company in Aberdeen had a Minor pick-up converted for the business. The cab is royal blue and the rear is white, sign written 'Bon Accord Glass, Aberdeen'.

Minor LCVs have been used for television and films, since they were first produced. They would often appear as working vehicles or as background for street scenes. More recently, they have been used in television advertisements. A Minor LCV appeared on an advert for 'Cadbury's Roses' in 1993 as a circus van, which came to the rescue of a broken down Volvo. The van appeared briefly in the scene. The registration number was GLK 774C. It might have been a 'Hi-Top' van, as the rear doors were extended and there was extra height in the rear back. A visit was arranged to 'Cadbury World' where the vehicle was seen. Alas, it was not a genuine 'Hi-Top', but a converted Post Office Telephones van. However, the registration number and the van were not quite right. Chris Stevens, at 'Post Office Morris Minors' (POMM), confirmed that the vehicle is ex-PO, but not 1965 as the registration implies. The body number, engine number, and general PO fittings date the vehicle as being 1971, not 1965. Also, there is no trace of green paint anywhere on the vehicle, only yellow, again indicating a much later date. The real registration is therefore unknown. The vehicle has had the rear-end cut along the waistline, including the rear doors and a twenty inch plate inserted and then welded up. It is not known if Cadburys carried out the conversion, or whether they purchased the vehicle like it. The vehicle carries a multi-colour livery of black, yellow and red. The van-back sides are black as is the interior. The front and rear doors and bonnet are yellow, the front wheel arches are red and the rear wheel arches are yellow. There are red and yellow stars all over the vehicle, with 'CIRCUS' written on the side panels, rear doors and over the cab. It also has two round windows fitted on either side panel, the roof has a large square cut out, with opening double. The fleet number should have been U114357and chassis Numbers OJE1/195325.

There is a television advertisement for National Westminster Bank, that begins with a train bursting out of the back doors of a van. It is sign written 'Supermodels'. This van belongs to Marcelle Burden's ex Oxford City Council lamp-lighting van, OJO 856J. It also starred in an American TV movie called 'Stick with me kid', starring Twiggy. The van was prepared as a delivery van for 'Kaufman's Snail Farm'. 'OJO' has also made two appearances in the BBC2 drama series 'Our Friends in the North', that was shown in April 1996, while also appearing in an ASDA commercial for their Easter shopping campaign as a hot cross delivery van.

'FPC 147J', which is a superb GPO van, owned by Brian & Joyce Banks, appeared on ITV in 1999 in Bill Bryson's 'Notes From A Small Island'.

Dave Thomas, the Register's founder used to have a vehicle, RRV 531L, that was a film star. It appeared as a Florist's van in 'Solitaire for Two'. It was filmed in Camden and Wimbledon, and the van was sign written 'Interflora Raymond's Florists'.

There are two minor vans in a video 'Madness, take it or leave it', by rock band Madness. One was yellow, with a red interior, registration number REU 777? The second is white van, registration number RWC 613K. Both feature throughout the eighty-two minute video and the yellow van is featured on the cover of the video. It was produced by *Virgin*, number VVD 114, during the early 1980s.

Ice cream van TNA 654K was used in the filming of *Titanic Town*, starring Julie Walters. It was filmed in Belfast, in 1999, and set in the late 1960s. The vehicle carried the registration number 7103 MZ. It has also been in the Granada film *East is East*, set in the 1970s, when it was JLN 826E.

Minors featured in two big films lately. *The Borrowers*, by Mary Norton, starring John Goodman, from 1997, filmed at Shepperton Studios, had four LCVs in it (three vans and a pick-up). It was used as an exterminator van, to eradicate the Borrowers. The van was converted for the film. A side panel was cut out and hinged back to house 'odd looking tools' and it was painted cream and orange. A green pick-up was used for what appeared to be a garage scene. The other two vans were used in a street scene where twenty-seven Minors are shown. One green van, complete with roof bars and sun visor, and a maroon one were used. All number plates were changed for the film.

Austin Powers, released in the UK in September 1997, was set in London during the 1960s. It was filmed at Paramount Studios, in Hollywood. A Minor pick-up featured with the registration number PKH 623A, for the film, owned by Rick Feilusch in the USA. Liz Hurley and Mike Myers starred.

Minor LCVs can be seen in films and on television programmes from the 1960s and 1970s, like *Randall & Hopkirk, Deceased*, *The Avengers*, *Department S* and *Man in a Suitcase*, to name a few. Most of these feature street scenes where Minor LCVs are seen. More recently, *Heartbeat*, on ITV, often featured Minor LCVs.

The Austin badge and crinkle grille was only found on the LCVs. It was produced from 1968 to 1971. *Photo: R. Harvey.*

Dave Thomas, wearing Panama hat, on the right, is the founder of the Minor LCV Register. The author is on the left. They are putting the banner up at the MMOC national show, Knebworth in 1994 *Photo: J. Harvey*

The Minor LCV stand and marquee at the annual Great Dorset Steam Fair, which is held annually at Blandford Forum. *Photo: R. Harvey.*

Minor LCVs at the Austin Morris Day, in 1997, from the other side of the 'Brooklands Hill climb', which is in between the white railings in the foreground. *Photo: LCV Register/ D Thomas Collection.*

This Morris van, FTD 219F, *c.*1968, was seen in Bullnose Road, Preston Docks, in 1998. *Photo: J. Baker.*

Mr Aresti's van, *c.*1971 (ex-BVA 941J, here in UK), is referred to as 'the silent salesman'. It is advertising his estate agency in Paphos in Cyprus. *Photo: N. Aresti.*

Bill Burton's Morris Series II quarter-ton van, 'RKH 819', was used in the filming of ITV's *Heartbeat* in 1997. The vehicle has trafficators mounted on extensions on the 'B' post. Lurking in the background, are the back doors of PC Bellamy's Mini police van, from the same programme. *Photo: B. Burton.*

This restored Morris van, 260 NTE, owned by Stan Pickles, is an early Series III, c.1957. He uses it for his business 'SCP Toupee Co' of Bolton. Stan made a hairpiece for Reg Holdsworth when he was in ITV's *Coronation Street! Photo: J. Murphy.*

This Morris 6-cwt van, 'RRV 531J', c.1971, was once owned by Southern Electric. Now it is owned by Dave Thomas. Here it is seen on the film set of *Solitaire for Two*, in 1996, starring Amanda Pays. *Photo: D. Thomas.*

Club member Jim Murphy is posing with the 'Sparks of Streatham' van which he restored to its former glory. This photo shows Jim being interviewed in Bolton Country Park on 23 June 1995, for the regional news programme *Granada Tonight*. *Photo: J. Murphy.*

Seen on the film set of *Hello Girls*, in 1996, is the Royal Mail van, *c*.1958, owned by Bob Logan. Colin Ellis's Morris J van is in the background. *Photo: C. Ellis.*

This yellow Post Office Telephones engineer's van, *c*.1971, is being used in the backdrop for a TV programme about GPO workers, in 1998. The van is owned by Colin Ellis. *Photo: C. Ellis.*

This circus van was converted from a PO Telephones engineer's van, c.1965. It is seen here on the film set for a commercial for *Cadburys Roses*, in 1993. *Photo: Haymarket Publishing.*

This is the film set of *Titanic Town*, starring Julie Walters, in 1998. It is set in Belfast, which is why Mr Seniuk's ice cream van 'TNA 654K', c.1971, is carrying an Irish registration number. This is the second back-end that Mr Senuick built himself over thirty-five years ago. It has been used on three of his vans. *Photo: D. Seniuk.*

A Morris Series II quarter-ton pick-up XKN 4, *c.*1954, restored to its former glory by its current owner, seen in 1999. *Photo: D. Markham.*

Morris 8-cwt van, UDB 950K, *c.*1971, has been wonderfully restored by Neil Ripley from Hull. It was adopted by the local 'Marie Curie Cancer Care' office during 1998 for promotions and is often used to advertise 'East Yorks Morris Minors'. *Photo: N. Ripley.*

In this re-created 1960s scene is Phoenix Furniture's working Minor USJ 311. It has a superbly written sign. The only give away is the telephone number and STD code shown on the van's side. *Photo: C. Douglas.*

This unusual angle shows off the sleek lines of the Morris Minor van. In the roof are the six holes, which would have held the ladder rack, as this is ex-PO Telephones engineer's van, KYX 957K, *c.*1971. Now Martin Smith owns it. *Photo: M. Smith.*

Contemporary by design? The car park is a stark background for this modified Minor, owned by Kevin Hollis. The modifications include the gleaming wolfrace wheels. *Photo: K. Hollis.*

What an advertisement for your business! This superb restoration, spotted in 1998, has now become a working van for 'Arkwrights Fish & Chips', on the Isle of Man. *Photo: R. Hunt.*

Bill Wilkinson poses outside the 'Sutherland Arms Hotel' in Scotland, in 1988, with a Morris Series III van, that advertises the hotel. *Photo: B. Wilkinson.*

Sarah Wright uses her Morris van to promote her parents' Millbrook Garden Centre, in Monmouth. The van carries the registration OMW 966, and is a 1961 Series III. *Photo: S. Wright.*

Nine

LCVs Overseas (Including scrap vans in the UK and abroad)

Minor LCVs, like the cars, were exported all over the world. Overseas plants dealt with completely knocked down ('CKD') models. Vehicles were packed in crates, in kit form, and reassembled wherever they went. Local suppliers were then used for parts and supplies such as tyres, glass, paint (Australian vans have different paint colours on the compliance plates) batteries and some interior and upholstery items.

It is not known how many Minor LCVs were assembled as part of the CKD kits, nor is it known exactly where, although Australia did assemble them.

Minor LCVs can be found in the USA, Canada, New Zealand, Holland, Scandinavia, Eire, South Africa. India, Pakistan, Germany, Austria, Greece and Spain. There are probably many others around the world rotting away in back gardens or under hedges, just waiting to be discovered and restored.

This chapter also covers vehicles that need a lot of 'tender loving care' and those under restoration.

Here is a Series II ex-Post Office mail-van, c.1955, in a scrap yard in West Wales, during 1979. *Photo: T. Bourne.*

Eight Morris Minor Series V vans are shown at a Danish Mail sorting office in 1969. The livery is yellow, with a black stripe around the van, at waist level, including the doors. All the postmen are wearing a very smart uniform of red jackets with gold buttons. *Photo: M. Wiidau.*

Danish Post Office van BL 60. 938 *c*.1960, described as 'Hillerød Postkontor', which refers to the post office in Hillerød. A wire mesh, as well as a wooden shelf for storing parcels, is visible behind the driver's seat. *Photo: Post oq Telemusum/Danish PO Museum.*

Wolfram Schonfelder, in Austria, with his Series II LHD van with its cheese grater grille, and side window conversion. The LCV is from 1954. *Photo: Series II Register Archives.*

Morris Series III pick-up LHD, *c.*1960, in St Helena, California, in 1977. It appears to be covered in the Californian dust! *Photo: K. Kernberger.*

This Morris Series III pick-up *c*.1959, has been superbly restored by Wilhelm Lombard in Worcester, South Africa. A Morris Minor saloon nestles to the right of this 1997 picture. *Photo: W. Lombard.*

Dieter Hedrich from Germany has restored his own Morris Series V pick-up LHD, *c*.1968. From this angle the later twin rear lights are visible. *Photo: D. Hedrich.*

This Morris quarter-ton Series III, *c*.1960, is from Nova Scotia. From this angle the small windows can be seen. The van has been superbly restored. It is advertising the *Village Green Motor Car Co. Ltd* and is fitted with chrome rear bumpers.

This Morris Series V, 6-cwt van, *c*.1970, was spotted in New Zealand, in 1972. It was being used as a support van for a rally car, travelling from one end of the island to the other. Peter Thompson, the owner, is shown with two admirers! Compare this with the previous picture, this one has larger rear windows. *Photos: P. Osbourne.*

This Morris 6-cwt van, 7909 ZX, has been restored and is used to advertise the *Town & Country Bar* in Bandon, Cork, 1997. *Photo: N. Ripley.*

Andres Sørensen, aged seventy-five, with his Morris Series III pick-up, c.1959. Both were going strong on 4 July 1999, in Denmark.

Raewyn Danks used her Morris 6-cwt van, pictured on the beach in New Zealand in 1997, for her lead glass business, hence the registration plate and sign on the side windows. The van used to be owned by the local council in New Zealand. *Photo: R. Danks.*

Stevan Dupus, from Cardiff Street, Kansas, in the USA, is the proud owner of this Morris 6-cwt pick-up. He has made some subtle modifications to the wheels, the rear lights and the rear chrome bumper. Photo: *S. Dupas.*

This 1960 Morris Series III pick-up, from Queensland, Australia, is shown in 1998. It is fitted with a chrome commercial bumper, which does not seem to be that unusual in Australia. *Photo: K. Harding.*

This Morris Series V van, c.1967, is believed to be the only Minor LCV in Luxembourg. It was imported from the UK in 1996. *Photo: B. Noesen.*

The rear box is certainly creased very badly, on this Morris quarter-ton Series II van, c.1956. The damage was as a result of a road traffic accident in 1962, in Denmark.

This is the same vehicle, which was repaired and repainted in 1962. All vehicles in Denmark had to be fitted with small indicators, as on the side of this van. The Saab in the background is also a rare vehicle today. *Photos: M. Wiidau.*

This Morris Minor pick-up, TF 62992, was spotted in Playa De Las Americas, Tenerife on 4 December 1999. The owner, Tony Diaz, uses it to promote his second-hand bookshop, 'The Book Swop'. It has rear chrome bumpers fitted from a traveller. *Photo: J. Baker.*

It's lunchtime for these mechanics who they shared with the owner of this Morris 6-cwt van Series V, c.1965, in Rawalpindi, Pakistan, in 1996. The van was undergoing a major brake cylinder repair. They were removed from the car to be bored out and re-sleeved, before re-fitting. *Photo: G. Murray.*

This is the same van, 'RIP 8926', as the one at the bottom of the previous page in Rawalpindi, Pakistan later that year. The owner is looking down from the *Café Rose* at the locals enjoying the nightlife. *Photo: G. Murray.*

Lans Hernes' Morris pick-up, *c.*1957, at Winter Camp 1999, in Norway. It is held every year in February and March, it is known as Primustreff, and has visitors from all over Norway, Denmark and Sweden. *Photo: L. Hernes.*

The Rocky Mountains form the backdrop for this Morris Series III pick-up, from 1960, owned by Bruce Blair, in Nevada, in the USA. In the distance is another LCV, also owned by Bruce. *Photo: B. Blair.*

A fleeting glimpse of the rear end of a Minor van in Kenya, in 1990. *Photo: K. Hollis.*

This ex-Post Office Telephones 6-cwt van, VNT 263J, lays abandoned in a field and it looks like it is beyond redemption! *Photo: LCV Register/Dave Thomas Collection.*

British Rail used to own this Austin Minor van. It was spotted a few years ago on a preserved railway line, presumably waiting to be restored one day! Photo: *LCV Register/Dave Thomas Collection.*

Acknowledgements

Firstly, sincere thanks to my father-in-law, Brian Lee, who gave me the confidence to attempt this book, for providing the contact with the publishers and for his guidance and advice throughout the production of this book.

For information, photographs, I should like to thank: AA Archives and Michael Passmore, B. Allen, R.J. Appleton, N. Aresti, B. Atkinson, J. Baker, A. Ballisat, R. Beardmore, K. Bennett, M. Bishop, M. Blaber, Bon Accord, T. Bourne, A. Browne, B. Bennett, M. Blackburn, B. Blair, J. Bowen, Bristol Water, L. Bourman, J.Brumwel, B.Carter,A.Chandler, C. Chipperfield, R. Crew, R. Crossman, S. Cook, Danish PO, R. Danks, C. De La Mare, C Dennett, J. Dicks, R. Doig, C. Douglas, S. Dupas, R. Edwards, C. Ellis, D. Estlea, E. Farrell, K. Farrell (Auto Express), L. Fleming, M Forster, D. Foster, R. Frost, D. & J. Game, M. Green (Jersey PO), T. Groves, I. Gutherson, J. Harvey, N. Harvey, P. Hanby, Haymarket Publishing, A. Hahn, K. Harding, D. Hedrich, K. Hollis, S. Holmes, B. Homans (PEVC), G. Hoskins, Hull Museum, R. Hunt, P. Jarman, B. Jenkins, J. Johnson, R. Johnson, K. Kornberger, Kingston Communications, G. Kinsman, N. Lawton, D. Leach, B. Lee, C. Letton, W. Lombard, M. Malloney, Marshalls, T. March, D. Markham, S. Marsbøll, Metropolitan Police Museum and Ray Seal, R. Montgomery, M. McGinty, G. Murray, P. Murray, J. Murphy, NRMA Archives, O. Naustdal, B. Noesen, P. Osbourne, H. Pehrsson, PM Photography, A. Parfitt, W.D. Pittham OBE, K.Porter, A. Probyn, RAA, RAC, M. Rapley, B. Read, B. Rhea, F. Rice-Oxley, N. Ripley, M. Roberts, J. Ross, B. Russell, SEB Museum, D. Senuick, B. Sharman, S. Shepperson, D. Short, M. Skillen, Skinner's Ice Creams (J Briggs), M. Smith, B. Stewardson, M. Street, Swedish PO, R. Taylor, R. Turner, J Vella, J. Wass, A. Watt, M. Wiidau, P. Willmet, J. B. Williams, P. Wood, D. Woodward, S. Woodward, S. Wright and C. Yarwood.

I should like to apologise to those whose photographs have remained unpublished, due to lack of space, and to anyone that I have inadvertently omitted who has contributed in any way to this publication.

Special thanks to 'Minor Monthly', 'Classic Car Weekly', 'Best of British', 'LCV News', 'Vintage Commercial Vehicles' and Morris Minor Owners' Clubs around the world, who printed requests for information and photographs. Thanks to Chris Stevens at the Post Office Morris Minors for his constant help with GPO & PO enquiries. A big thank you to Bill Wilkinson, Dave Thomas and Eric Payne who have endured numerous telephone calls, in my quest for information, as well as providing photographs.

Lastly, thanks to my good friend Godfrey Crew, not only for the photographs, but for the constant get-togethers and cups of tea, as he reminisced about Minor LCVs.

And finally... This is the author's van, which is currently under restoration. The previous and only owner was Thomas Hewertson, a wholesale tobacconist in Newport, South Wales. *Photo: R Harvey.*